P9-CBR-677

CHEMISTRY MAGIC

# CHEMISTRY MAGIC

Kenneth M. Swezey

*McGRAW-HILL BOOK COMPANY, INC.*

NEW YORK TORONTO LONDON

**CHEMISTRY MAGIC**



*Library of Congress Catalog Card Number: 56-10332*

Second Printing

**McGraw-Hill Text-Film Department** is now distributing 39 science films featuring experiments described in Mr. Swezey's earlier books: AFTER-DINNER SCIENCE and SCIENCE MAGIC. Planned with the junior-high school science curriculum in mind, these experiments use equipment and materials readily available in the classroom or at home.

While maintaining a sound educational approach, these films combine high visual interest and clear explanations, bringing science subjects to life. Each film deals with one basic scientific principle, making that principle more understandable by relating it to experiences in everyday life.

The films cover material found in such units as:

WORK AND MACHINES  ENERGY FROM THE SUN  THE AIR AROUND US
COMMON CHEMICAL AND PHYSICAL CHANGES  MAGNETIC AND ELECTRICAL ENERGY

The Junior Science Films may be obtained from the Text-Film Department, McGraw-Hill Book Company, 330 West 42d Street, New York 36, New York. Each film runs 13 minutes and is priced at $55.

*Published by the McGraw-Hill Book Company, Inc.*
*Printed in the United States of America*

# A Word to the Reader

HERE is a handbook of easy-to-do experiments that have been worked out especially for four groups of chemistry enthusiasts:

• The home laboratory hobbyist who wants to continue beyond the smell-making and color-changing stunts of the conventional juvenile chemistry book.

• The teen-age chemistry-set fan who is anxious to graduate to more elaborate experiments performed with professional equipment.

• The student chemist, chemistry club leader, and chemistry teacher who needs a source book of demonstrations which show how chemistry is applied in everyday life and industry.

• The chemistry student who would like to start a home laboratory but doesn't know how.

The pattern of *Chemistry Magic* was conceived in the Home Chemistry department of *Popular Science Monthly,* which the author conducted for a dozen years. From among thousands of chemistry-fan readers came letters from teachers, students, hobbyists: "We already know how to make carbon dioxide from vinegar and baking soda and how to change water into wine; please tell us how to make bakelite, rayon, and synthetic dyes!" "How do chemists make food safe and purify water?" "What is meant by catalysts and polymerization?" "Chemical reactions are all very interesting, but what do they mean to everyday life?"

Apparently there were legions of amateur chemists who had outgrown the parlor-trick variety of chemical magic and were now eager to explore the far greater magic of the chemistry that creates new textiles and plastics, tests food and catches criminals, contributes in a thousand ways to the progress and comfort of mankind. Teachers, too, evidently needed experiments in applied chemistry to supplement the pure chemistry demonstrations of their textbooks.

The experiments that follow are part of a series chosen and worked up by the author to meet this new need. Many were devised originally for *Popular Science*; others appear here for the first time.

Demonstrations in *Chemistry Magic* range all the way from how to make bakelite, rayon, and synthetic rubber to how to remove spots from your clothes. Chapters on atomic energy, paper chromatography, and sleuthing with ultraviolet light introduce the reader to latest techniques. Where a subject itself is familiar, the author has tried to devise new experiments or to associate older ones more closely with everyday affairs. For the beginner, several introductory chapters explain the basic tools and procedures of the chemist and tell how to get started.

Each chapter or experiment unit consists of half a dozen or more experiments related to a single subject and is complete in itself. Although some of the earlier units may be a little easier than later ones, each is independent and may be undertaken in any order desired.

Because standard laboratory equipment costs little or no more than the homemade variety it has been used throughout on the theory that the home chemist will get more fun and satisfaction from his hobby by working with the tools of the school and professional chemist.

Contrary to popular notion, the dangers involved in chemistry experimenting are no more mysterious or sinister than those dealt with in daily living. From early childhood we all have to learn to keep gasoline away from fire, to avoid swallowing iodine or other household items labeled POISON, to keep from spilling drain cleaner or other corrosive substances on hands or clothes. The dangers of the chemical laboratory are no different from these and must be avoided in the same way—by the exercise of knowledge, care, and common sense.

Most of the experiments in *Chemistry Magic* are harmless, no matter how you do them. In the few experiments that involve chemicals or operations that might prove dangerous if not properly handled, specific instructions are given as to how to conduct the experiments safely. General rules for the safe handling of dangerous chemicals are also given in the chapter on basic laboratory technique. If the experimenter faithfully follows these cautions and directions he should experience no trouble whatever.

The author wishes to thank the editors of *Popular Science Monthly* for permission to reprint material that originally appeared in its pages. Thanks, too, to George Lord of the Fisher Scientific Company for help concerning apparatus and chemicals.

My gratitude goes also to the following faithful friends who suffered long hours under hot lights posing for the illustrations: Raymond Albert, Daniel Byrne, Donald Cluen, Joseph Curley, William Curry, Richard Dempsey, Eugene Duffy, David Findlay, William Halpin, Frank Kelly, Thomas Killeen, Daniel Korbelak, Frank Lynch, Peter McAllister, William McGarry, Wayne Overton, Kenneth Quigley, and Joseph Roe.

*Kenneth M. Swezey*

# Contents

## CHEMISTRY IN INDUSTRY



## CHEMISTRY IN THE HOME

# CHEMISTRY MAGIC

A siphon bottle makes a handy water supply for your lab. Once started, water will flow every time the pinchcock is released.

# The Chemist's Workshop

IF you are a lucky amateur chemist, your laboratory may be the well-appointed "lab" of a local school or chemistry club, or it may occupy a substantial part of a cellar, attic, or spare room in your own home. If you are not so fortunate, however, you can still do real chemical work with just a cabinet for your chemicals and apparatus, and a table on which to set them up while experimenting. The difference in result is chiefly a matter of atmosphere and convenience.

A room that can be set apart from outsiders, especially inquisitive youngsters, makes an ideal place for your experimenting. Spilled chemicals will do less damage to the floor if it is of bare wood or concrete. At least one window is necessary to provide ventilation when experimenting with irritating or flammable fumes and gases. If possible, provision should be made for running water and gas.

**Laboratory furniture.** Often old household equipment can be adapted to serve as furniture in a home lab. An ancient marble-top dresser or chest of drawers of convenient height, or an old kitchen cabinet base with a plastic or linoleum top, makes an excellent work table. You can store apparatus in the drawers. Wood or lineoleum tops can be made chemical-resistant by the application of several coats of special laboratory paint, obtainable from a chemical supply house.

Apparatus and harmless chemicals may be stored in old bookcases or other furniture provided with shelves. Poisons and combustible chemicals, however, should be kept in steel cabinets (floor- or wall-type kitchen cabinets will do) which can be locked if there is danger of unauthorized tampering. *(Even with this precaution, home chemists should never store more than the minimum necessary amount of any dangerous chemical!)* Provide a 1-gal earthenware crock, or a metal wastebasket painted inside with several coats of laboratory paint, in which to dispose of waste chemicals, soiled filter paper, and broken glassware during experimenting.

**Siphon bottle.** Lacking a sink near your table, a homemade siphon bottle will provide water for your experiments. Make this from a gallon jug, a 2-hole stopper, two pieces of glass tubing, rubber tubing, and a pinchcock, assembled as shown in the photo on the opposite page. The glass delivery tube should nearly touch the bottom of the jug and the rubber tube should reach several inches below the bottom.

Mount the filled bottle on a shelf a foot or so above your table. To start the siphon, loosen the pinchcock and tilt the bottle until water starts to flow from the rubber tube. Then close the pinchcock. Water will thereafter flow from the tube each time the pinchcock is opened until the bottle runs dry.

**Tools to work with.** If you are an experienced home or school experimenter, the everyday tools of the professional chemist are already familiar. If you are a beginner, or a chemistry-set enthusiast who wants the further thrill and satisfaction of working with standard equipment, however, the following paragraphs will help introduce you to the basic apparatus you will need. Less common equipment will be described in the experiments in which it is required.

Buy apparatus only as you need it. Names and addresses of laboratory supply houses may be found in advertisements in school chemistry and popular science magazines.

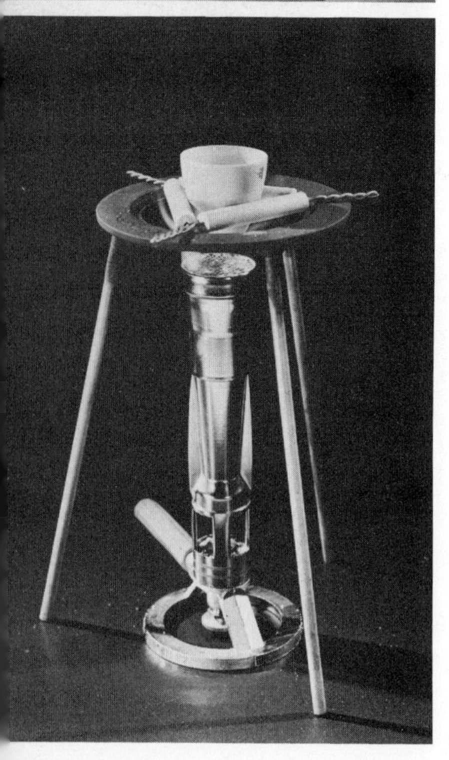

**Supports for apparatus.** A ring stand is probably the handiest single support you can own. You can buy it alone, or with three or four clamp-on rings which can be used to hold a funnel or to support a flask, beaker, or crucible over a flame. Fixed and adjustable clamps may be added which will steady the neck of a flask, support a test tube, burette, or thermometer, or perform an assortment of other holding jobs. A ring stand with a base at least 5 by 8 in. and a rod 20 in. high will be more useful than a smaller one.

*A tripod* may substitute for a ring stand for supporting a beaker, crucible, or evaporating dish at a fixed height over a burner. This three-legged iron stand may have a single ring for a top or it may have several concentric rings which can be removed or added to suit the size of the apparatus to be heated.

**Sources of heat.** If you have a supply of gas in your lab, a bunsen or a Fisher burner will solve practically all your heat problems. If you have not, you will have to improvise. A small alcohol lamp will generally do for heating test tubes. A larger alcohol or gasoline burner, or an electric hot plate (preferably one with the heating element enclosed) will heat flasks and beakers. A propane soldering torch, provided with a flame

How a ring stand (left, above) is set up to support a flask over a bunsen burner. Place a wire screen with asbestos center between ring and flask to distribute the heat. How a tripod and triangle (left) support a porcelain crucible over a Fisher burner.

spreader, will give you plenty of heat for working glass tubing. You can effectively dry chemicals and evaporate solutions in an evaporating dish by directing the rays from a 250-watt reflector-spot or infra-red bulb down on them.

*The bunsen burner* is versatile and inexpensive, and may be bought in types for either natural or artificial gas. A wing-top attachment will spread the flame for glassworking. The hottest part of a bunsen flame is a point just inside the tip of its outer cone.

To light a bunsen burner, close the air supply at the base of the burner, turn on the gas at the source, and bring a lighted match just above the mouth of the burner from one side. Then gradually open the air supply until the original luminous yellow flame has turned nonluminous and blue. To regulate the height of the flame, turn the gas supply up or down and readjust the air supply. If the flame pops down the tube, or "strikes back," it was probably getting too much air. Turn off the gas and start over again.

*The Fisher burner* is an improved modern form of air-gas burner which costs more than the bunsen type but which gives a broader flame and considerably more heat. Unlike a bunsen burner, the Fisher burner is hottest about ½ in. above the grid, right in the heart of the flame.

Support test tubes diagonally, with the mouth away from you (right, above). Boiling is smoother if tube is heated a little above the bottom. Chemicals may be dried effectively under the rays of an infared or 250-watt reflector spotlamp (right).

**Florence flask, Erlenmeyer flask, beaker, and test tubes: these are the basic glass cooking and mixing pots of the laboratory.**

For best results with a Fisher burner, open wide the stopcock to the gas supply and ordinarily open wide the air ring on the burner. After lighting, adjust the flame with the needle valve at the base of the burner until it tends to leave the grid. The flame is then at its optimum temperature.

**Laboratory glassware.** Few experiments are performed without the assistance of flasks, beakers, or test tubes. Before using any of these, be sure they are sound.

*Test tubes* are ideal for heating or reacting small quantities of chemicals. They may be supported by an adjustable clamp on a ring stand, or by a wire test-tube holder held in the hand. Support them diagonally, with the mouth away from you and anything or anybody else who might be injured if the contents should be ejected. To prevent "bumping"—the violent eruption of large bubbles of steam caused by superheating of the lower part of a liquid—heat liquids cautiously, starting near the top and working down. Leave the flame finally a slight distance above the bottom of the tube. For heating dry chemicals, or for the destructive distillation of coal, wood, and the like, use test tubes of Pyrex or other heat-resistant glass.

*Beakers* may be used for heating larger quantities of chemicals. Always protect these from direct flame by means of a square of wire gauze with an asbestos center. When heating or boiling a liquid for a very short period, a beaker may be filled about three-quarters full. For boiling a solution away to dryness, however, fill the beaker only one-third full and cover it to prevent loss from spattering.

*Florence flasks* are the type generally used for boiling or distilling. Liquids are less likely to bump in the round-bottom than in the flat-bottom type. A gauze square should be used under flasks as well as beakers.

*Erlenmeyer flasks* are used, in preference to beakers, for dissolving and mixing chemicals which must be protected from unnecessary contact with the air and other chemicals. Mixing can generally be accomplished by gently swirling the flask. These flasks are also used for long boiling operations, when it is desired to retain as much of the solution as possible. The tapered walls help condense and return the vaporized liquid.

*Thistle tubes* are usually straight glass tubes with a thistle-shape funnel at the upper end. Generally inserted through one hole of a 2-hole stopper, they are used as a means to add liquid to gas generators or other closed flasks.

*Pipettes* are narrow glass tubes, plain or graduated, used for measuring, transferring, or dropping small amounts of fluids. The common "medicine dropper" is a simple pipette of the dropping type.

*Wash bottle.* Every laboratory should have at least one wash bottle. Distilled water from its nozzle can be made to wash down solid matter from the walls of beakers and test tubes, to moisten filters, and to rinse the inside of glassware before drying. You can easily make this piece of apparatus from a 500-ml flask, a 2-hole stopper, a short length of rubber tubing, and three pieces of glass tubing. Bend two of the glass tubes as shown in the photo at upper right. The glass part of a medicine dropper will make a good nozzle, or you can draw one from a short length of tubing.

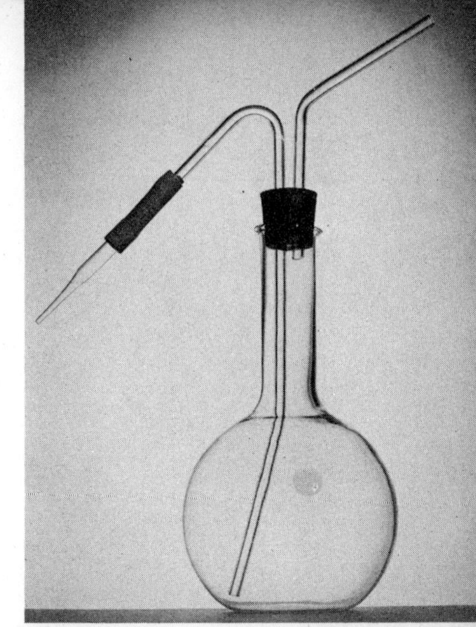

A handy wash bottle (top) can be made from a flask and tubing. A crucible and an evaporating dish (center and bottom) are porcelain cooking dishes of the chemist.

**Porcelain crucibles** are used instead of test tubes and beakers for work at high temperatures that would crack or melt glass. Use them for melting metals and preparing alloys, for burning off residues, and for chemical reactions producing great heat. To exclude oxygen and to increase the heating effect, they may be covered. Crucibles may be heated directly in a flame and are generally supported by a pipestem-covered triangle of wire on a tripod or ring stand.

*Porcelain evaporating dishes* are used for evaporating solutions to a small volume or dryness, and for baking. Except in the very smallest sizes, they should be heated on a wire gauze and never over an open flame. Heating solid chemicals or evaporating to dryness in an evaporating dish more than about 150 mm in diameter may cause it to break.

*A mortar and pestle* of porcelain is handy for grinding coarse chemicals to a powder. Because porcelain is brittle, though hard, never attempt to pulverize substances that require heavy pounding in this equipment.

**Tools for measuring** are needed in almost every step of chemical experimenting. Here is the minimum: graduated cylinders to measure the volume of liquids, a scale to measure weight, and a thermometer to measure temperature.

*Graduated cylinders* with straight sides are easier to read than the traditional graduate of the pharmacist and so are preferred by the chemist. One of 25 ml and another of 250-ml capacity will do for a beginning. To read the volume of a liquid in a cylinder, hold the cylinder so that the bottom of the curved surface of the liquid is at eye level. The graduation then directly in your line of sight indicates the correct volume.

*Scales.* A photographer's balance costing less than $10 will serve for most home chemistry experiments, including all those in this book. Be sure you get a model with *metric* weights. With this, you can weigh up to 60 g with an accuracy of about 0.1 g.

*Thermometers* for general laboratory use come in different ranges. The lowest reads from $-20°$ to $+110°$ C, and the highest from $-10°$ to $+500°$ C. The price (as well as the length) of a thermometer goes up with the temperature range. For economy and general usefulness, one reading from $-20°$ to $+150°$ C is a good buy.

**Odds and ends.** Besides the basic apparatus already mentioned, the home chemist will need a few supplementary items before he can start work. Here are the most important.

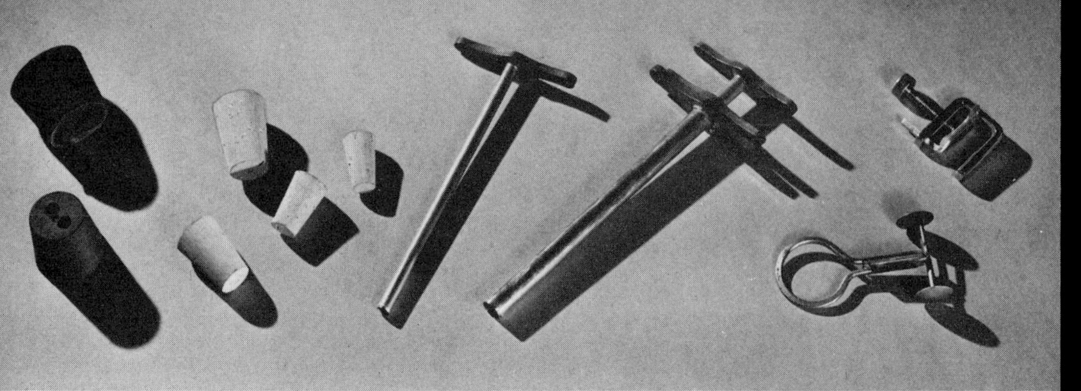

*Rubber stoppers* come solid, with one hole, and with two holes. Sizes 2 to 6 are those most often used. To economize, you can get an assortment of 2-hole stoppers and close unused holes with short lengths of glass rod, cut off and fire-polished as explained on page 14.

*Corks* are cheaper than rubber stoppers and are just as good for many purposes around the lab. In cases where solvents may attack rubber, corks are actually better. The "velvet" grade is a smooth cork of best quality and is recommended for use with volatile liquids and in distillation. Next comes XXXX and XXX, in descending order.

*Cork borers* for cutting holes in corks and rubber stoppers are tubes of thin brass or steel with a cutting edge at one end and a handle at the other. In common sizes, they come in sets of three or more. The steel variety costs more than the brass but keeps its edge longer.

*Glass tubing* for connecting apparatus generally comes in 4-ft lengths and in sizes from 3 mm to about 50 mm in diameter. Soft glass tubing of 6 mm or 7 mm diameter will probably be the most useful.

*Rubber tubing* used for temporarily joining glass tubing should be strong, flexible, and have an inside diameter slightly smaller than the tubes it joins. If solvent vapors or corrosive gases are to pass through the combination, make the joint short and bring the glass ends nearly together. Always use strong, thick-wall rubber tubing for making the connection to your gas burner.

*Pinchcocks* are clamps, operated by spring or screw pressure, for closing off or regulating the flow of a liquid or gas through a rubber tube. For partially closing a tube, or for closing a heavy tube completely, a screw clamp should be used.

*Brushes* for cleaning chemical glassware come in all sizes and shapes. You should have at least a small brush that will get inside test tubes and a larger one to scrub out flasks and beakers. Ordinary wire-core, stiff-bristled household brushes will do. Clean small glass tubing with an extended variety of "pipe cleaner," available from lab supply houses in 10-ft lengths.

For safety's sake, hold test tubes and other equipment well away from you during a chemical reaction. Watch progress from the side.

# Basic Laboratory Technique

THERE is almost no limit to the number of experiments which may be performed with simple equipment and only a limited number of operations. If you familiarize yourself thoroughly with the use of chemicals and apparatus and master the best technique for carrying out various operations, your experimentation becomes safer, simpler, and more scientifically valuable.

The following hints are fundamental to all laboratory practice, and should be kept always in mind when working out experiments in your home or school laboratory.

**Cleanliness.** To begin with, the chemical laboratory and all operations conducted in it must be kept scrupulously clean. A tiny speck of foreign chemical in a bottle of chemically pure reagent may spoil a test. A few

grains of chemicals mixed in error may cause a fire or an explosion. For these reasons, never return excess chemicals to a reagent bottle. Either save the chemicals separately for another and less critical experiment or throw them away. Small quantities of dry reagent may be removed from a bottle with a clean spatula or a strip of clean paper folded lengthwise to form a scoop. Small quantities of liquids may be obtained with a clean medicine dropper or pipette. Never lay glass stoppers on a table. Either hold them or place them in a clean glass or porcelain dish.

When weighing chemicals, never place them directly on the scale pan. Use a creased square of white paper under the chemical, balancing this with a similar square on the other pan. Have a supply of these papers on hand and throw them away as they are used.

On precision scales, weights should always be handled with forceps to prevent possible contamination or corrosion of them by substances from the fingers.

**General rules for safety.** • Wear an apron, a smock, or old clothes when experimenting. Spattered chemicals will quickly ruin good clothing.

• Never mix chemicals haphazardly—just to see what will happen! You won't learn anything from such a procedure and you stand a good chance of causing a dangerous accident.

• Use the *smallest* amounts of chemicals necessary to get the desired result. To use more is wasteful and a confession of poor technique. In the case of poisons and explosives, the use of more than minimum amounts may be definitely dangerous.

• Perform experiments at arm's length. Tilt the mouth of test tubes in which you are heating or reacting chemicals away from you. Don't look into chemical equipment during a reaction.

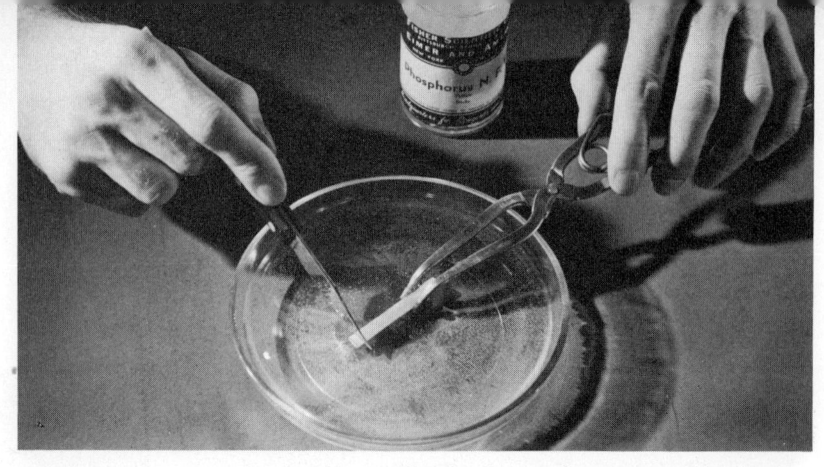

As phosphorus catches fire if exposed to air, it must be stored and cut under water. It must never be handled except with tongs.

• Smell gases and other odors cautiously. First waft a little to your nose with a cupped hand to test its identity and strength. A hearty whiff of ammonia, chlorine, sulfur dioxide, or any of a number of other corrosive gases will send you reeling and may possibly injure your eyes and throat.

• Have plenty of ventilation when experimenting with poison gases. If you are not lucky enough to be able to operate your apparatus under a ventilating hood, perform such experiments outdoors or near an open window.

• Never taste chemicals you are not thoroughly familiar with. Some are poisonous in extremely small amounts.

• When diluting acids—especially sulfuric—*always add the acid slowly to the water,* with constant stirring. Adding water to acid may cause a violent reaction that may spatter you with hot acid.

**Safety with dangerous chemicals.** Observe those caution signs on bottles of chemicals! Keep bottles containing acids or ammonia away from your face when opening them. Rinse and dry the outside of acid bottles before returning them to the shelf. Bottles marked POISON should be handled with special care, and the contents should be kept away from the mouth and skin. Wash your hands repeatedly while using poisons, and carefully clean up any spilled chemicals with rags that can be thrown away. If these precautions are observed you need not fear to use poisons.

A few chemicals require extra-special handling. Yellow phosphorus, for example, must be kept under water, as it catches fire spontaneously in the air. Never touch this chemical with your fingers. You may be seriously burned. To cut a piece for use, remove a stick of it from its bottle with tongs and place it immediately in a dish filled with water.

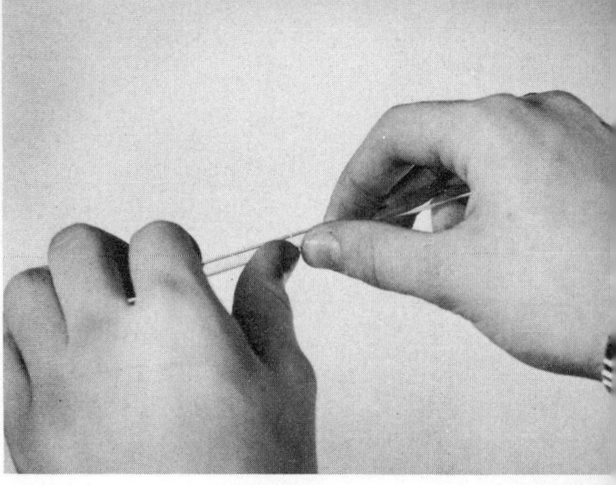

To cut glass tubing, first make a sharp scratch across it with a file. Then hold it as shown above and bend the ends toward you.

While still holding it with the tongs, cut pieces off the other end with a sharp knife. Be sure all phosphorus used for an experiment is either burned or returned to the bottle. Pieces left around may cause a fire.

Sodium and potassium metals must be handled with similar care. These, however, must be kept under kerosene, as water causes them to decompose violently and liberate hydrogen, which often catches fire from the heat of the reaction. Cut these in a dish under kerosene. Scrape off any coating of oxide and dip the pieces in a little ether to remove the kerosene. When reacting or cutting these metals, it is best to wear goggles to protect your eyes from possible spattering. Never throw waste sodium or potassium down a drain, as the reaction with water might cause a serious explosion. Small scraps may be destroyed by adding them slowly to denatured alcohol in a beaker. After the scraps have disappeared, this alcohol may be bottled and specially labeled, to be used subsequently for the same purpose.

That KEEP FROM FLAME warning found on certain containers means you must keep them at least *twenty feet* from any exposed flame—especially when opening them. The vapors of some flammable liquids travel far and fast. Even if they do not carry a flame back to the bottle, they may form an explosive mixture with the air.

**Simple glassworking.** Any home chemist can quickly become expert in cutting, bending, and otherwise shaping glass rods and tubes up to about 12 or 14 mm in diameter. Ordinary soft glass can be worked fairly well in the flame of a bunsen burner. Pyrex glass requires the hotter flame of a Fisher burner.

To cut a small-diameter tube or rod, lay it on a flat surface and make a single sharp scratch across it with a triangular file. Then hold

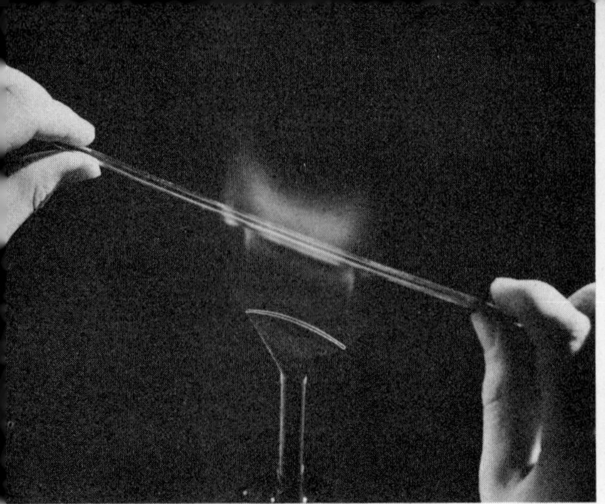

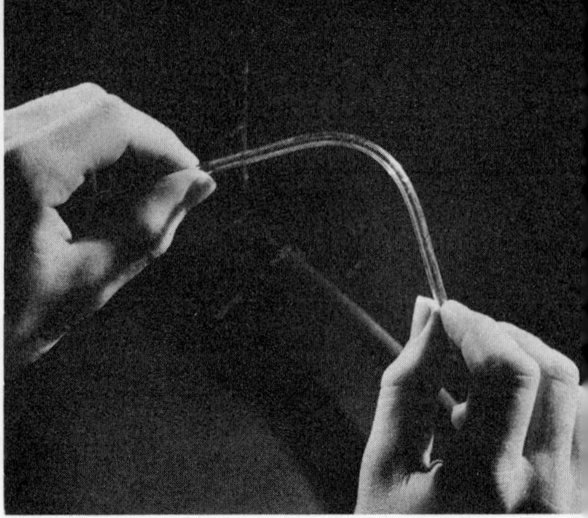

**To bend glass tubing, first rotate it in a bunsen flame until the flame turns yellow. Then remove it and make your bend.**

the tube with both hands, the tips of your thumbs touching each other and the tube directly opposite the scratch. If you now bend the ends of the tube toward you, and pull at the same time, the tube should break off neatly. To cut larger tubes, you may have to make a deeper scratch that extends entirely around the tube.

To prevent cutting your hands and damaging rubber stoppers, the jagged ends of tubes and rods should always be *fire-polished*. This is done by rapidly rotating the end, just inside the topmost cone of a bunsen flame until the glass starts to soften and flow. A tube should ordinarily be removed from the flame before the hole starts to diminish in size. If you want to seal the end, however, just continue to rotate it in the flame until the hole closes.

A jet may be drawn in a piece of tubing by first holding both ends and rotating the center slowly in a bunsen flame until it becomes red hot. Then remove the tube from the flame and draw the ends straight apart until the central part is about 2 mm in diameter. When the tube is cool, cut the narrow part as described above, and fire-polish the tip.

To make perfectly round bends without kinks, a sufficient length of tubing is first heated as shown in the photo. Use a wing top on your bunsen burner to spread the flame. Hold the tube loosely and rotate it slowly to heat it evenly on all sides. Remove the tube from the burner when the flame turns yellow. Then, holding the ends with your fingers, bend the tube gently to the proper curvature.

**Glass tubes and rubber stoppers.** Cuts from the jagged edges of glass tubing broken while attempting to insert or remove the tubing from a hole in a rubber stopper are by far the commonest casualty in the home and school laboratory. They can be avoided by using care.

Before trying to insert a tube in a stopper, be sure the rubber of the stopper is pliable, the hole not too small, and the end of the tube is fire-polished. Moisten the tube and the inside of the hole with water. Then, protecting your hand with a towel, grip the tube close to the stopper and push it into the hole with a gentle twisting motion. Never use great force or hold the tube in such a way that, should it break, the jagged end would penetrate the palm of your hand. If the tube requires too much effort to insert, substitute a smaller tube or enlarge the hole with a cork borer.

Unless the apparatus is to be permanent, always remove tubing from stoppers immediately after use. Otherwise the removal may become difficult and dangerous. If a rubber stopper has firmly hardened around a tube, remove it by slitting one side lengthwise with a sharp knife and then peeling it off. It is better to sacrifice a stopper than to risk breaking the tube and cutting yourself.

**Filtering.** This is a laboratory operation which can be speeded and improved if proper attention is paid to details. To make a filter, first fold a disk of filter paper into halves and then into quarters. Open the folded paper to form a cone and insert this in a funnel large enough to extend slightly higher than the cone. Moisten the paper with water from your wash bottle and press it carefully against the sides of the funnel.

The remainder of the filtering set-up is important, too. Arrange the long point at the end of the funnel so that it touches the inside of the receiving beaker near its top. This permits the flow of the filtrate down the side of the beaker without splashing.

Liquids may be filtered without dripping or splashing by means of the arrangement shown in the top photo. After washing and rinsing chemical glassware with tap water, rinse it twice with distilled water from your wash bottle, as shown above.

Hold a glass stirring rod across the top of the pouring beaker and allow the liquid gently to strike the side of the filter paper about one-fourth of the way down from the top. This glass-rod arrangement is useful whenever liquids are poured from beakers. If the beaker has no lip, the edge should be lightly greased with petroleum jelly to prevent the liquid from running down the side.

**Laboratory dishwashing.** Beakers, flasks, and other chemical glassware should be washed in the simplest way possible. Water solutions of acids, alkalis, and salts usually wash out easily with plain tap water. Ordinary dishwashing detergents will generally remove an oily film deposited by the fingers or petrolatum. More stubborn grease may be removed by soaking the glassware in a 5 per cent solution of trisodium phosphate, followed by a scrubbing with a stiff brush. Some organic deposits may require the application of an organic solvent, such as alcohol, gasoline, benzene, or carbon tetrachloride.

After washing, the glassware should be rinsed several times in clear tap water and then twice with distilled water sprayed from the wash bottle. Invert the glassware to dry on a rack or drainboard. Never dry the inside with a towel, because this leaves lint and perhaps chemical contamination.

**Disposing of waste chemicals.** Mention has already been made of methods for disposing of scrap sodium, potassium, and phosphorus. Calcium carbide also needs special attention, as it generates flammable and explosive acetylene gas when in contact with water. Let your waste calcium carbide react completely with water, outdoors or in a well-ventilated room, before discarding it.

It is probably unnecessary to warn home chemists not to pour strong acids and other corrosive chemicals down the drain of the kitchen sink. If you live in the city, dilute such chemicals with a large amount of water and flush them down the toilet. Small amounts of less active chemicals and solid chemicals may be emptied into an earthen crock, or metal pail protected inside with several coats of acid-resisting paint, to be disposed of in the same way at the end of each experiment session. If you live in the country, waste chemicals may be buried in a special pit far from your garden and water supply.

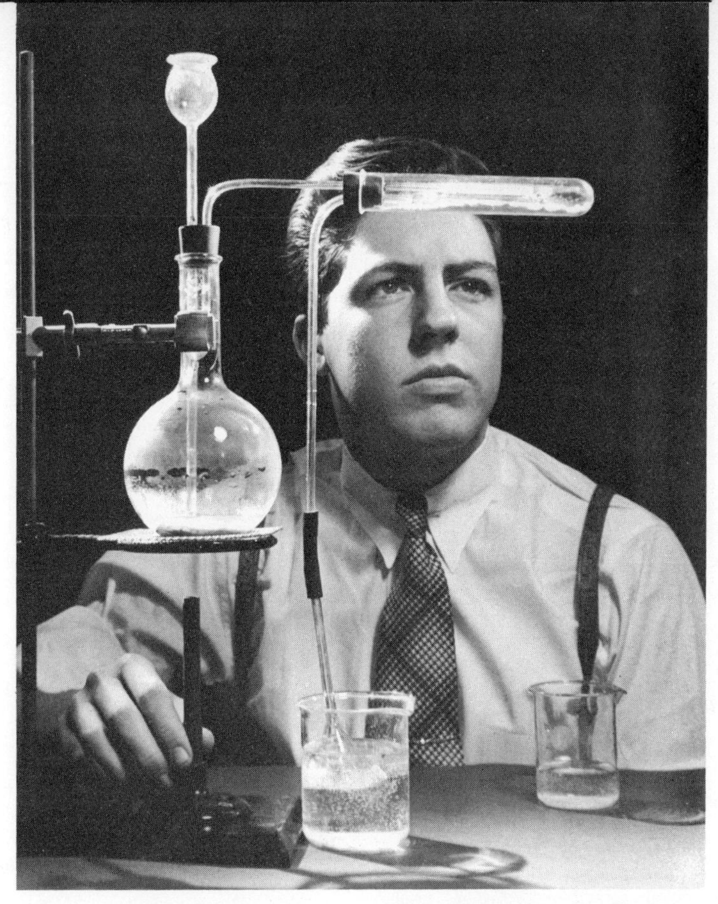

You can make calcium hypochlorite, the popular "chloride of lime" bleach, by passing chlorine gas over slaked lime in a test tube.

# Calcium—the Builder's Element

FEW people have ever seen the silver-white metallic element calcium, for it never occurs free in nature and its commercial application is extremely limited. Its compounds, however, are among the most abundant and important chemical elements on earth.

Mortar made of slaked lime, or calcium hydroxide, was used more than 5,000 years ago to build the pyramids of Egypt just as it is used today. Limestone and marble, sea shells and coral reefs, the great chalk cliffs of Dover, crystal caves, double-refracting Iceland spar—all are forms of calcium carbonate. Calcium hypochlorite is the well-known "chloride of lime." Gypsum and plaster of paris are calcium sulfate. Largely as material for teeth and bones, calcium phosphate makes up about 1.5 per cent of your body weight.

Precipitated chalk is formed by mixing a warm dilute solution of sodium carbonate with a similar solution of calcium chloride, as shown in the top photo. Cold, concentrated solutions of these chemicals form a jelly, as at right, above. Stirring liquefies the mass and precipitates chalk.

Sir Humphry Davy named calcium from the Latin *calx,* which means chalk. Chalk cliffs and beds are composed chiefly of the compacted skeletons of microscopic marine animals. Powdered chalk is the "whiting" used as a filler in rubber and other products, as an abrasive, and as a pigment extender for paints. Precipitated chalk, the principal ingredient in most tooth powders and pastes, is produced by means of chemical reaction.

**Precipitated chalk.** With the help of any soluble calcium salt and any soluble carbonate, you can demonstrate how chalk is precipitated. Calcium chloride and sodium carbonate are the chemicals generally used commercially. Make a warm dilute solution of each and pour one into the other. The white dense cloud that forms is finely divided calcium carbonate. Allowed to stand undisturbed, the chalk will settle to the bottom.

**Solid changes to liquid.** Substitute cold, concentrated solutions of calcium chloride and sodium carbonate for warm, dilute ones, and you can perform a curious feat of chemical magic. Mix these two with gentle but thorough stirring, and the result is a jellylike solid. Continue stirring for a few minutes, and the mass turns into a milky liquid. Let it stand a few minutes longer, and the white coloring material settles to the bottom once more as precipitated chalk.

**How lime is made.** Limestone, one of the most abundant forms of calcium carbonate, is the source of the millions of tons of "quicklime" (calcium oxide) and "slaked lime" (calcium hydroxide) needed yearly for mortar, plaster, cement, and the neutralizing of acid soils. The limestone is broken and heated to high temperature in huge furnaces or kilns until carbon dioxide is driven off. The resulting quicklime is crushed to the size desired, and screened.

You can show how this process works with the help of a chip of limestone or marble and a bunsen flame. Twist a piece of iron wire around the chip and hold it in the hottest part of the flame for several minutes. Then dip it into water containing a drop of phenolphthalein solution. The chip turns pink and colors the water, indicating that the originally neutral stone has become an alkali. By driving off carbon dioxide, the flame changed the original calcium carbonate into calcium oxide; the water, in turn, changed the calcium oxide to calcium hydroxide.

**Slaked lime.** The last compound, cheapest of all alkalis, is made by carefully adding water to quicklime until visible reaction stops. Pure calcium oxide will combine with about 30 per cent of its own weight of water. Considerable heat and steam are produced in the process, and finally the calcium hydroxide collapses as a fine powder. Mortar is made by mixing slaked lime and water with sharp sand. Its hardening is caused by a very slow chemical change in which the carbon dioxide of the air transforms the hydroxide into hard calcium carbonate.

**Limewater,** made by shaking a little slaked lime in water and filtering off the undissolved lime, makes a useful test for carbon dioxide. Blow into the clear solution through

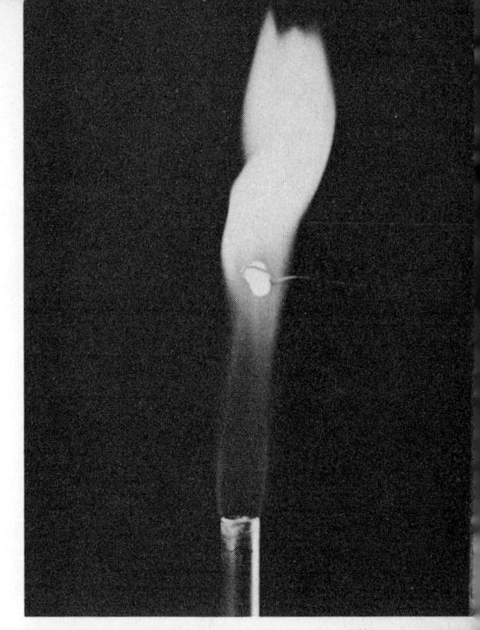

Heating in a flame drives carbon dioxide from a chip of limestone or marble, leaving calcium oxide (quicklime). Dipped in water containing phenolphthalein solution, the calcium oxide changes to the alkali calcium hydroxide and turns the solution pink.

Most calcium compounds can be formed by adding calcium carbonate to a solution of the desired acid until bubbling stops and then evaporating the water. Here calcium chloride is being made.

a tube. It becomes milky as insoluble calcium carbonate is formed. Continue to blow, however, and the solution clears again because excess carbon dioxide changes the carbonate into soluble calcium *bi*carbonate.

**Hard water and stalactites.** The fact that calcium carbonate dissolves in water containing excess carbon dioxide accounts for giant limestone caves, the stalactites and stalagmites that "grow" in them, and certain types of "hard" water. Water containing carbon dioxide from the air and running through fissures in limestone rocks dissolves minute portions of the rocks. After centuries, the fissures grow into tunnels and then into caves. Icicle-like stalactites of stone hanging from the roofs of such caves are caused by the seeping through of water containing calcium bicarbonate and the subsequent evaporation of the water, leaving the solid carbonate behind. The upside-down "icicles" on the floor, the stalagmites, are produced by the accumulation of calcium carbonate from the drippings of the stalactites above them.

**How to make calcium chloride.** Calcium carbonate is the starting point for most of the calcium compounds. Sprinkle precipitated chalk into a dilute solution of the desired acid until carbon dioxide stops bubbling, and you can make any calcium salt you need. With hydrochloric acid, the salt in solution is calcium chloride. Evaporate the water and the compound forms needlelike crystals, which when further heated in an evaporating dish become dry calcium chloride, a chemical with an amazing affinity for water. This dehydrated calcium chloride is used in the laboratory to dry gases, and in damp cellars and closets to remove excess moisture from the air. It is sprinkled on roads, tennis courts, and in the subways, to lay dust by keeping the earth damp with absorbed moisture.

**Plaster of paris.** Vast deposits of calcium sulfate occur in nature as gypsum—a *di*hydrate in which two molecules of water are combined with each molecule of sulfate. When gypsum is carefully heated to between 120° and 180° C, three-fourths of the water of crystallization is driven out, leaving a *hemi*hydrate. This is "plaster of paris," made first of gypsum from the slopes of Montmartre. The "setting" of plaster of paris is a real chemical change, in which the partly dehydrated crystals of calcium sulfate regain the water they lost during the heating. The resulting hard mass is produced by the interlocking of myriads of microscopic crystals.

**Household "chloride of lime,"** despite its name, is vastly different from the calcium chloride previously mentioned. It is technically calcium hypochlorite, a compound made by passing chlorine gas over thin layers of slaked lime.

You can easily make a sample of this common chemical by means of the setup shown on page 17. Put 10 g of granular manganese dioxide in the flask and add 25 ml of concentrated hydrochloric acid. Spread a thin layer of calcium hydroxide along the side of the horizontal test tube. Arrange a small inverted funnel so that its mouth barely dips below the surface of a dilute solution of sodium hydroxide (ordinary lye will do) in the beaker. This latter will absorb most of the excess chlorine and prevent the gas from contaminating the air in the room.

Heat the acid in the flask very gently, never allowing it to boil. Greenish chlorine gas soon pours into the test tube. Part reacts with the lime, while the remainder is carried into the sodium hydroxide solution below. After several minutes of operation you will have a calcium hypochlorite powder as effective as the household disinfectant and bleach.

Magnesium is such an active metal it will burn even in steam.

# Magnesium—the Bantamweight Metal

AS a lightweight structural metal for aircraft parts, and as a pyrotechnic material for star shells, signal flares, tracer bullets, and flash and incendiary bombs, magnesium leaped into fame during World War II. Strong, silvery-white, and only two-thirds as heavy as aluminum, magnesium is the lightest of all construction metals. In the form of powder, thin sheets, or wire, it burns with a dazzling white flame that water or even carbon dioxide will not put out.

Never found alone in nature, magnesium is made on a tremendous scale by the electrolysis of its compounds. These compounds are among the most plentiful substances in the crust of the earth. Whole mountain ranges are made up of dolomite, a double carbonate of magnesium and calcium. Asbestos, talc, and meerschaum are magnesium silicates. Epsom salts, named after the springs at Epsom, England, where they were first isolated in 1695, are magnesium sulfate. In the form of its chloride, there are nearly 6 million tons of magnesium in every cubic mile of the sea.

**Compounds of magnesium.** Less spectacular, perhaps, than the metal, the compounds of magnesium are just as important. Asbestos and magnesium oxide are among our most valuable insulators against heat. Magnesium oxychloride forms a superior artificial stone and flooring material. Magnesium carbonate is used for insulation and for making dentifrices, talcum powder, other magnesium chemicals, and Pyrex glass. Epsom salts, citrate of magnesia, and milk of magnesia are three of the many magnesium compounds that are employed in medicine.

With only a box of epsom salts as a starting material, you can make many of these compounds in your home or school laboratory. With a few inches of magnesium ribbon, you can likewise test some of the exciting properties of the metal itself.

**Magnesium burns in steam.** One amazing property that helps make magnesium metal such an important incendiary material for wartime use is its ability to steal oxygen from such ordinarily stable compounds as water and carbon dioxide. During the war, magnesium fires generally were extinguished by smothering with sand. Water helped only when applied in quantities sufficient to cool the metal below the point of combustion.

As a demonstration of magnesium as an oxygen grabber, boil some water in a flask, and then with tongs lower a short length of lighted magnesium ribbon into the steam. Instead of going out, the magnesium continues to burn brightly, getting oxygen by decomposing the steam.

**Carbon dioxide,** usually one of the best fire-extinguishing materials, is as helpless as steam against burning magnesium. Fill a beaker with this gas by pouring ½ in. of water into it and then adding a little baking soda (sodium bicarbonate) and vinegar or any other common acid. As soon as the bubbling has stopped, test for carbon dioxide by lowering a lighted match into the beaker. The match will go out at once.

Now lower a piece of burning magnesium into the glass and see the difference. The metal goes on burning furiously. In the process, it

Burning magnesium robs oxygen from carbon dioxide, as at left, and unites with inert nitrogen, as the experiment above shows.

changes into magnesium oxide, and black specks of carbon, wrested from the carbon dioxide gas, are flung to the sides and bottom of the beaker.

**Magnesium joins with nitrogen.** Even nitrogen, which ordinarily is one of the most inert gases, will unite with hot magnesium when conditions are right, forming magnesium nitride. To show this, put a few short pieces of magnesium ribbon on the center of an upturned can cover and heat the cover over a gas flame until the metal catches fire. Then remove the cover from the flame and allow it to cool until you can touch it with your hand. Now put several drops of water on the warm substance that remains and hold a bit of cotton wool moistened with hydrochloric acid above it. White telltale smoke of ammonium chloride immediately arises. On burning, the magnesium united with oxygen and nitrogen from the air, forming the oxide and the nitride. When water was added, the nitride decomposed into ammonia gas and magnesium hydroxide.

**Most magnesium compounds** can be produced from the carbonate or the hydroxide. The carbonate occurs naturally as magnesite and, mixed with calcium carbonate, in certain forms of marble and limestone. It can be made artificially by mixing hot solutions of magnesium sulfate (epsom salts) and sodium carbonate (washing soda). Since the carbonate is not soluble in water, it is precipitated as a fine white powder. When dried, it can be used as a polishing agent and for heat insulation.

**Magnesium hydroxide** also is made by precipitation. Again you can start with your epsom salts, this time adding a solution of sodium hydroxide (or ordinary lye). When dissolved, both of these solids produce clear solutions. Mixed together, however, they form a white gelatinous precipitate. A suspension of this, in pure form, is the "milk of magnesia" of the drugstore.

**Magnesium oxide.** By strongly heating either your hydroxide or carbonate, you produce magnesium oxide (magnesia), used widely for heat insulating, the lining of high-temperature furnaces, and for making oxychloride cement.

**Magnesium chloride.** By dissolving either the carbonate or hydroxide in hydrochloric acid you get magnesium chloride, the compound found in sea water from which magnesium metal is made in vast quantities. To obtain crystals of this chemical, evaporate the solution over a water bath until nearly dry. Then complete the drying in a warm place in the open

Mixed, water-clear solutions of lye (sodium hydroxide) and Epsom salts (magnesium sulfate) form a white gelatinous precipitate, magnesium hydroxide. In purer form, this is the basis for the antacid "milk of magnesia."

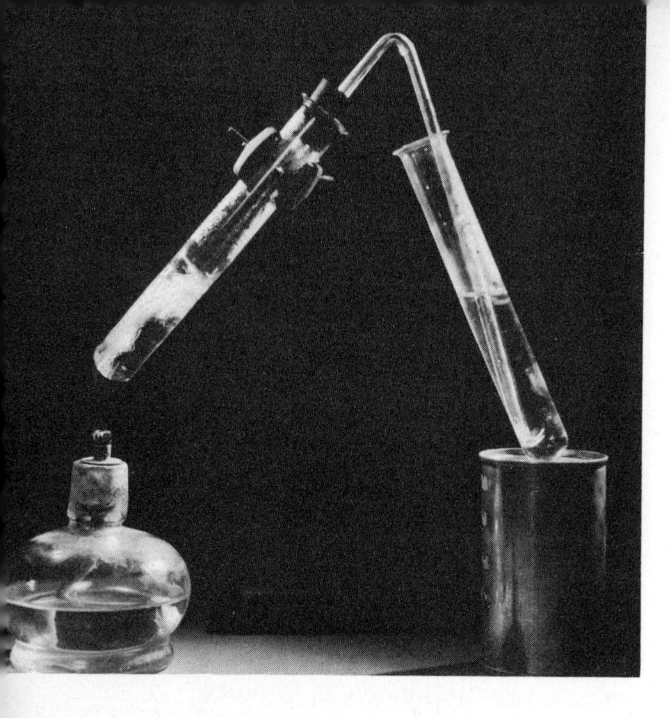

You can make hydrochloric acid by heating magnesium chloride crystals until they melt and boil and allowing the vapor produced to bubble through water. Part of the vapor is the gas hydrogen chloride. Dissolved in water, this forms hydrochloric acid.

air. If the heating is continued too long, the crystals will lose some of their water content and partly decompose, giving up hydrochloric acid.

**Hydrochloric acid.** This decomposition of magnesium chloride by heat provides one method for the manufacture of hydrochloric acid, if other sources should fail. As a demonstration, put some crystals of magnesium chloride in a test tube and fit the tube with a stopper having a glass delivery tube long enough to reach the bottom of a second test tube, which is half filled with water. Now gently heat the crystals. They will first melt, and then vapors will pass through the delivery tube and bubble up through the water. Part will be water vapor and part hydrogen chloride. The latter will dissolve in the water, changing it into hydrochloric acid.

**Cement from magnesium.** By dissolving as much magnesium chloride as possible in water at room temperature, and then mixing into this solution enough magnesium oxide to make a thick paste, a cement is formed which reacts rapidly to produce a hard durable material used widely for imitation stone and for the stonelike flooring of office buildings. Called "Sorel cement," after its inventor, the material is also often referred to as "magnesium oxychloride cement," although its actual composition may vary because of different conditions of mixing and setting.

When ignited, chromic oxide and aluminum powder react violently to leave molten chromium that hardens into a button of pure metal.

# Chromium—Tough Guy among Metals

BRIGHT, silvery-white metallic chromium is almost as hard as a diamond and as tarnishproof as gold. Deposited electrically over iron and steel, pure chromium has almost entirely displaced nickel as a wear-resisting decorative coating for household appliances, automobile trim, and machine parts. Because of their extremely hard surface, chromium-plated gears, gauges, dies, and printing plates are tougher than those of unplated casehardened steel.

Mixed in small quantities with steel, chromium increases the strength and resistance to corrosion of the steel without increasing its brittleness, though chromium itself is very brittle and can be broken quite easily. Safes, projectiles, automobile and airplane engines, high-speed tool steels, big guns, and battleship armor plate all contain from 1 to 2 per cent

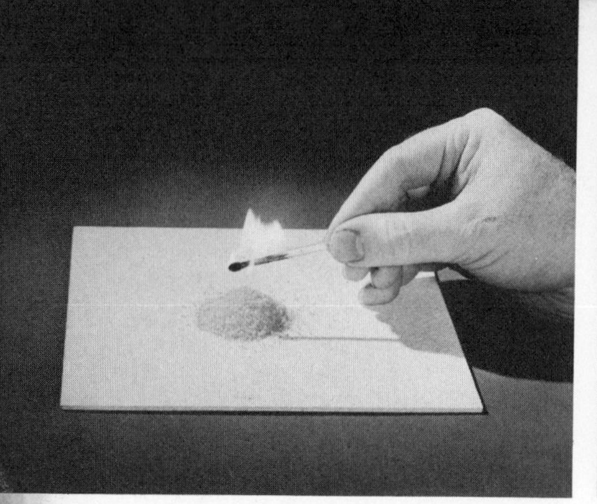

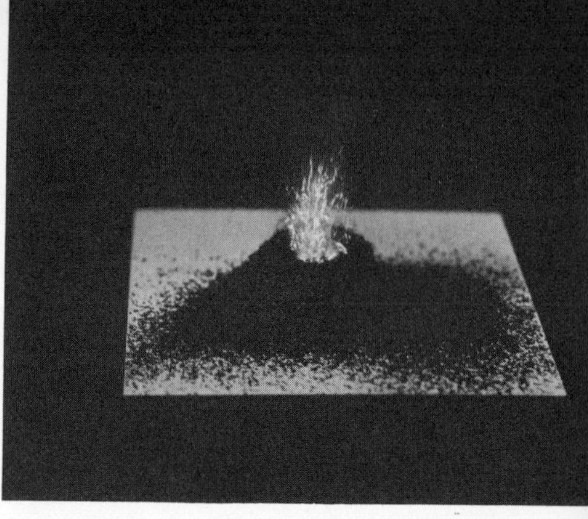

chromium. Rustless iron and stainless steel contain about 14 per cent. Nichrome, an alloy of chromium with nickel, is widely used in the form of wire as a resistance element for electric heaters.

**Chromium's colorful compounds.** It was not the silvery metal but the compounds of chromium that gave this element its name. The word "chromium" is derived from the Greek *chroma,* which means "color." Compounds of chromium range through every color in the spectrum. The green of the emerald and the red of the ruby are both due to traces of chromium. Chromic compounds are generally violet or green. Chromous compounds are blue. Lead chromates, used as pigments and dyes, are yellow and orange. Bichromates, also known as "dichromates," are usually orange-red.

**A chromium volcano.** With half an ounce of ammonium bichromate you can perform a novel experiment having a surprise ending. This experiment also introduces you to several colors of the chromium rainbow and provides you with a chemical from which you can make chromium metal.

Heap the ammonium bichromate in the form of a low cone on the center of an asbestos mat or the bottom of an inverted cake tin. Then light the top of the cone with a match. As soon as the chemical starts to burn continuously, turn out the room lights. Your little mound now looks like an erupting volcano, shooting forth red sparks from its crater and giving off a voluminous ash that resembles lava. When the eruption ends, turn on the lights again. In place of the little heap of orange-red compound, you'll find a mound of olive-green chromic oxide. The burning is really a decomposition in which nitrogen and water are broken free from the ammonium bichromate, leaving a puffed-up residue of chromic oxide.

**This chromic oxide green** is a permanent, nonfading pigment used for coloring glass, enamels, pottery, and the like. It is not the more fugitive color known simply as "chrome green" or "Brunswick green," which is formed by combining prussian blue with chrome yellow and is used as a pigment in oils, printing inks, and textile prints.

**Chromium metal made from oxide.** By combining finely granulated aluminum with part of your chromic oxide and igniting it, you can perform an experiment more spectacular than the first. In this experiment, aluminum plunders oxygen from the chromic compound—and at the same time raises the temperature to more than 2,000° C and releases a button of molten chromium. Pure chromium is produced commercially by this same reaction, called "Goldschmidt's process of aluminothermy."

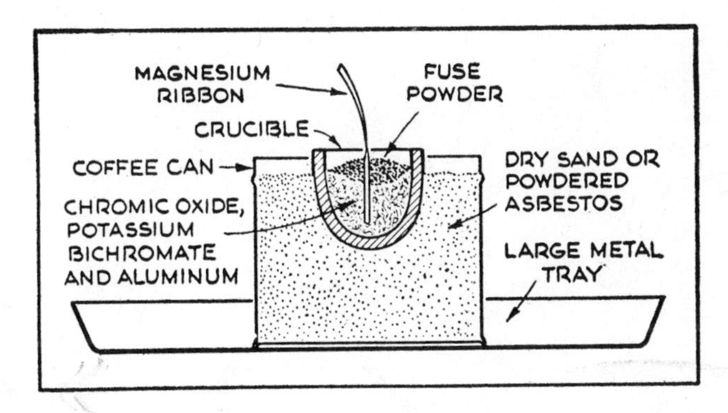


This experiment is not dangerous if you limit the chemicals to very small quantities, perform it outdoors or in a room with a bare concrete or wood floor, and keep out of the range of sparks, which fly three or four feet. Protect your eyes also with dark goggles or a sheet of smoked glass so you can watch the intense glow with safety.

**For a melting pot** use a small, replaceable fire-clay or porcelain crucible. To conserve heat for the reaction, as well as for the sake of safety, embed this crucible to its rim in a coffee can filled with powdered asbestos or dry sand. Set this in a larger metal tray or pan to catch near-by particles.

You'll find that you can squeeze the chromic oxide you made in the previous experiment to much less than its fluffed-up volume. This done, heat it to redness in an iron pan for a few minutes to drive out moisture. Then melt 2 g of potassium bichromate in another pan, allow it to cool,

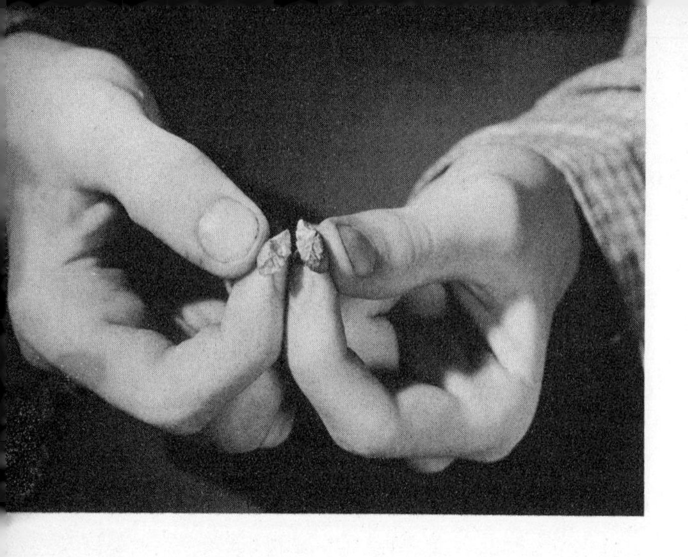

Chromium metal is so brittle it may be broken with pliers, but it is so hard that a sharp edge will scratch glass or steel. It is silvery and resists corrosion.

and powder it in a mortar. Weigh out 10 g of chromic oxide, add this to the potassium bichromate, and in addition put in 2 g of finely granulated aluminum (coarse powder). Mix these three substances thoroughly and pack them in your crucible.

**A fuse powder** is needed to bring this mixture to a temperature high enough to start the reaction. This can be made by mixing ½ g of powdered aluminum with ½ g of potassium perchlorate. Now make a little depression in the center of the chromium mixture, pour in most of the fuse powder, and spread the remainder over the top of the mixture. Then insert a 4-in. length of magnesium ribbon, pushing it through the powder to the bottom of the crucible.

**To start the reaction,** carefully light the end of the magnesium ribbon and retreat quickly to a distance of at least four or five feet. The effect is most startling in a darkened room. Amid intense light, heat, smoke, and fireworks, the aluminum robs your green chromic oxide and red potassium bichromate of oxygen, freeing the chromium from both compounds.

When the crucible has finally cooled, break it open and remove the button of chromium. You will find this metal brittle enough to break with pliers, but a sharp corner of the break will be hard enough to put scratches on glass or ordinary steel.

**Potassium chrome alum,** one of the commonest chromium compounds, is a double sulfate of potassium and chromium. It has exactly the same formula as ordinary potassium alum, but with chromium substituted

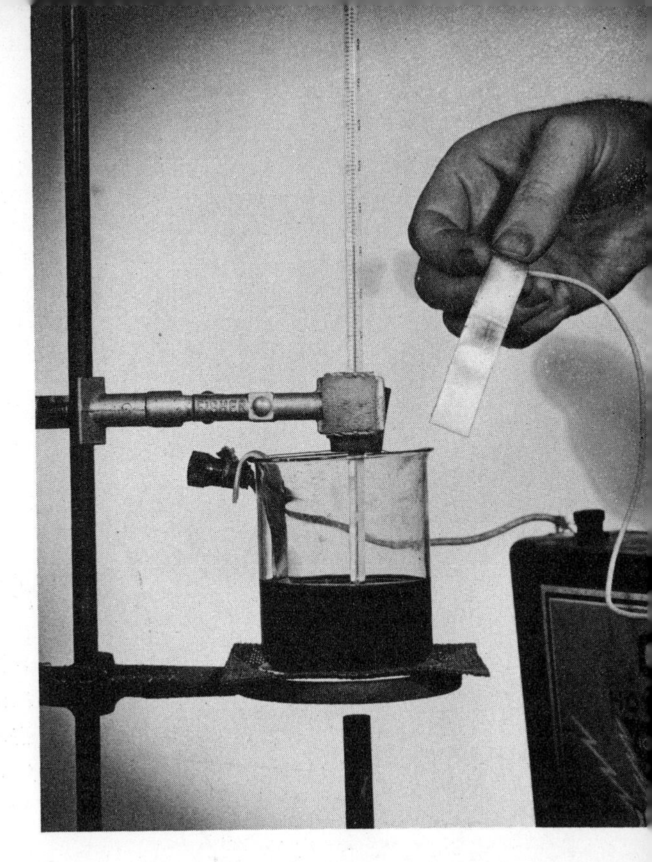

Chromium can be plated on metals by electrolysis. The nature of the plating depends on the temperature and concentration of the bath and the current strength.

for aluminum. It is used in tanning, as a hardener in photography, and as a mordant for dyes.

**You can make chromium hydroxide** by dissolving some potassium chrome alum in cold water to form a violet solution and adding ammonium hydroxide. A green-gray precipitate of hydroxide is formed which may be washed by decantation and dried. When this green-gray hydroxide is heated strongly, green chromic oxide is the result.

**Potassium bichromate,** an orange-red crystalline salt, is probably the most used of the chromium compounds and the one from which most of the other compounds are made. It shares with other bichromates the feature of being capable of changing to a yellow chromate when in solution merely by the addition of a base. Dissolve some potassium bichromate in water and add, with stirring, a little ammonium hydroxide. Suddenly the change in color indicates the change in chemical composition. Likewise, by adding a little mineral acid (hydrochloric or sulfuric) to the yellow chromate solution, it will be quickly changed back to orange-red bichromate.

**Chromium plating.** The development of satisfactory methods of plating with chromium from solutions of chromium trioxide has revolutionized the whole plating industry within the last few decades. All the best methods are patented and cannot be used commercially without license, but you can easily demonstrate the principle in your home lab. You will need some chromium trioxide (chromic acid), concentrated sulfuric acid, a 6-volt battery, and a 1- by 3-in. lead plate as an anode.

For your plating solution, dissolve in a 250-ml beaker 40 g of the red chromium trioxide crystals in 100 ml of water. *(Caution: Be careful in handling this substance. It is a powerful oxidizing agent and reacts destructively with many organic materials, including cloth, paper, and skin. If you spill any on yourself, rinse immediately with plenty of water.)* To this solution add 12 drops of concentrated sulfuric acid.

Set up the beaker on a ring stand over a bunsen burner and support a thermometer to reach well into the solution. Bend one end of the lead anode, hook it over the side of the beaker, and connect it with a wire to the positive pole of your battery. Bring the temperature of your bath to about 45° C before you try to plate anything, and keep it there during the plating.

**The character of chromium plating**—whether it turns out to be dull or bright, or whether it takes place at all—depends upon careful control of the strength of the bath, the temperature of the bath, and the strength of the electric current. A bright plating should be obtained on smooth, clean metal surfaces with the concentration and temperature specified and a current of 1 ampere per square inch of surface being plated.

Plating is most easily done on bright copper or nickel surfaces, which should be connected by a wire to the negative pole of the battery and lowered into the bath. The current may be regulated somewhat by moving the object being plated closer to the lead anode or farther from it, or a small radio rheostat may be connected in series with the wire from the object and the battery.

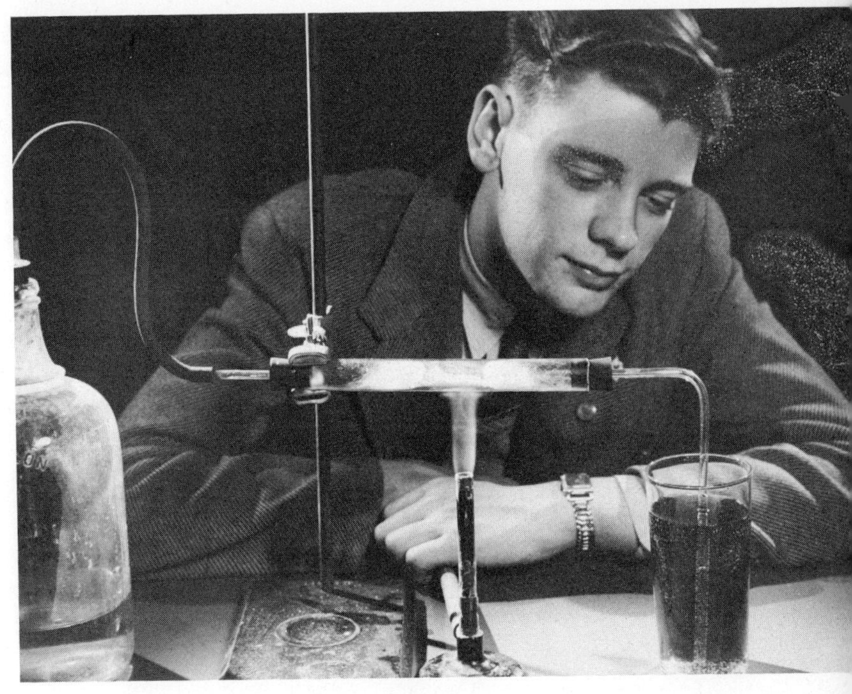

Cinnabar, heated in the horizontal tube, breaks up into mercury and sulfur dioxide. The latter decolorizes solution at right.

# Mercury—the Liquid Metal

MERCURY, the only metal that is liquid at ordinary temperatures, is one of the most famous and fascinating of the elements. Because of its wide distribution and the simplicity of its metallurgy, it was known to the ancients. It was the *prima materia,* or "prime material," which the alchemists believed could be changed into gold, silver, and other elements. Since then it has found wide use in medicine and in the arts.

Both the common term "quicksilver" and the Latin chemical name *hydrargyrum,* "water of silver," eloquently describe the elusive liquid metal that does not wet glass and that is so heavy that iron nuts, bolts, and washers float on it like corks on water. Because of its weight, mercury is an ideal liquid for barometers and suction pumps. Its high boiling point (357° C) and uniform expansion to heat make it an excellent fluid for thermometers. Although mercury actually can be changed to gold now by the miracle of atom smashing, the transmutation will never make anyone rich, for the cost far exceeds the value of the final product.

Most of the world's mercury is obtained from cinnabar, a naturally occurring red sulfide. By roasting this ore in the presence of air, mercury vapor and sulfur dioxide gas are driven out. The mercury is condensed into liquid form by cooling, and may then be refined by means of distillation.

**Make mercury from cinnabar.** You can easily demonstrate how this is done. Mount a tube of hard glass, about an inch in diameter and ten inches long, as shown in the photo on page 33. With the help of a folded strip of paper, carefully place a little powdered cinnabar or mercuric sulfide in the center of the tube, and keep it there with a loose plug of asbestos or glass wool on each side.

Connect one end of the tube to an outlet tube in the stopper of a gallon bottle that is also connected, through another tube, to a water supply. Plug the other end with a stopper fitted with a bent glass tube. Insert this bent tube in a tumbler of water containing several drops of sulfuric acid and enough potassium permanganate to color it a pale violet.

After making sure all connections are tight, heat the center of the tube strongly. Then allow a small stream of water to run into the bottle. This forces a slow current of air through the roasting tube, causing the generated gases to bubble up through the solution in the tumbler. The solution will lose color slowly, revealing that one of the decomposition products of heated mercuric sulfide is sulfur dioxide. The other, metallic mercury, condenses in tiny globules in the cooler parts of the roasting tube.

**Metallic mercury is not poisonous,** but its vapor is quite poisonous. Mercury therefore should never be heated in an open vessel. All soluble mercury salts are also poisonous, and they should be handled with care. Keep them away from your mouth and food, clean up immediately any you spill, and wash your hands thoroughly after handling them.

**Familiar compounds of mercury** are insoluble mercur*ous* chloride, or "calomel"—used largely as a purgative—and its deadly companion, mercur*ic* chloride, better known as "corrosive sublimate" or "bichloride of mercury." Despite its potential deadliness, mercuric chloride may be also a lifesaver as a powerful antiseptic. Because light can decompose calomel into dangerous corrosive sublimate, this medicine must be kept in dark bottles.

**White of egg acts as an antidote.** The white of an egg, mixed with a little water and administered quickly, is one of the best first-aid antidotes for mercury poisoning. Mercury salts precipitate albumin, forming a bland, insoluble substance that takes the remaining salt "out of circulation." To demonstrate the reaction in a test tube, pour diluted egg white into a solution of mercuric chloride or nitrate. A hard, insoluble precipitate will form and remove the mercury from solution.

**Mercury forms compounds easily.** Mercury has the strange property of uniting directly with most metals and a number of other elements to form compounds and alloys. Grind a little metallic mercury with powdered sulfur in a mortar, for instance, and the substances soon unite to form black mercuric sulfide. Iodine crystals ground with mercury produce red mercuric iodide.

**An alloy of mercury and another metal** is called an "amalgam." The "silver penny" that boys make by rubbing mercury on a clean bronze cent is the result of amalgamation. Zinc rods of wet primary batteries are coated with mercury to provide an alloy surface that acts as if it were pure zinc. Plastic amalgams of mercury and silver or gold are used as dental fillings, their plasticity being controlled by the proportion of mercury. Surfaces of metal objects may be amalgamated with mercury also by dipping them in a solution of a mercury salt and then rinsing them with water.

Dipped in mercuric chloride solution, a strip of polished copper is quickly coated with metallic mercury as shown in the top photo. Mercury is so heavy that steel nuts, bolts, and washers float on it like corks, as shown above.

**Mercury chemical makes writhing snakes.** One of the most spectacular salts of mercury is mercuric thiocyanate, the chief ingredient of the famous "Pharaoh's serpents' eggs." When one of these "eggs" is lighted, it produces a voluminous ash which curls grotesquely like a live snake.

You can make mercuric thiocyanate by slowly pouring a saturated solution of potassium thiocynate into a saturated solution of mercuric nitrate until the formation of a white precipitate stops. A few drops of a solution of ferric chloride in the original mercuric nitrate solution will show better when the reaction is completed. Stir this mixture constantly while adding the potassium thiocyanate, and stop pouring when it turns a stable pinkish hue.

Allow the solution to stand for half an hour, then filter it and wash the precipitate on the filter paper three or four times with cold water sprayed on it from a wash bottle. Next remove the filter paper from the funnel, unfold it, and lay it in a warm place to dry. Don't use artificial heat to hasten the drying or the substance might ignite spontaneously.

When dry, put the powder into a mortar or glass dish and work it into a stiff paste by adding a little water containing a few drops of mucilage and a few grains of potassium nitrate. Then mold the paste into conical pellets, about a quarter of an inch in diameter and half an inch high, and place them on a sheet of glass to dry. *(Caution: Mercuric thiocyanate is poisonous, so wash your hands thoroughly after handling it.)*

To change a pellet into a writhing reptile, place it on an asbestos pad or an inverted pie tin and light the tip. It will burn with a blue flame, slowly giving off as it does so a wriggling snakelike ash fifty or more times as long as itself.

If iron is put in a solution of copper sulfate and sulfuric acid, copper will be deposited. This is due to a swapping of metals.

# The Chemical Activity of Metals

EVERY home chemist knows from experience that different metals react in widely different degrees with other chemicals. When a small piece of potassium or sodium is placed in plain water, the metal reacts violently with the water, releasing hydrogen so furiously that the gas often catches fire. Magnesium won't release hydrogen from cold water, but it will from hot water. Aluminum, zinc, iron, and tin can't displace hydrogen from water, but can from acids. Copper, silver, platinum, and gold won't set hydrogen free even from acids.

By elaborate tests, chemists have arranged all metals in a list according to the ease with which they enter into chemical reactions. This list, shown on the next page, is variously called the "activity series," the "electromotive series," the "electrochemical series," or the "displacement series," depending upon the use for which it is intended.

## THE ACTIVITY SERIES

| | |
|---|---|
| [illegible] | [illegible]ium |
| K | potassium |
| Ca | calcium |
| Na | sodium |
| Mg | magnesium |
| Al | aluminum |
| Zn | zinc |
| Cr | chromium |
| Fe | iron |
| Ni | nickel |
| Sn | tin |
| Pb | lead |
| H | HYDROGEN |
| Bi | bismuth |
| Cu | copper |
| Hg | mercury |
| Ag | silver |
| Au | gold |

Lithium, the most active metal, stands at the top; gold, the least active, at the bottom. Hydrogen acts much the same as a metal in displacement reactions and so is included as a guidepost. Any metal above hydrogen is more active than hydrogen, and so will displace this gas from such acids as sulfuric and hydrochloric. Metals below hydrogen cannot displace it from any of the acids.

**See how metals react.** To impress this on your mind, you can make a visual demonstration. Add one part of hydrochloric acid to four parts of water and pour an equal amount of the diluted acid into each of four test tubes. Into one tube drop a piece of freshly polished copper, into the next a similar piece of iron, into the third a piece of zinc, and into the last a bit of magnesium. The magnesium reacts so strongly that the acid literally boils. Zinc releases bubbles a little less rapidly, iron displaces the gas rather slowly, copper not at all.

Magnesium is so active chemically that it will release hydrogen from such weak acids as lemon juice and vinegar. It will even do so from boiling water.

**One metal ousts another.** Another principle to be learned from the activity series is this: When any metal is placed in a solution of a salt of a metal that stands below it, the first metal is dissolved and the second is thrown out of solution. You can prove this in one instance by means of a color change. Dissolve some copper sulfate in water, and the copper ion will color the solution blue. Now add a little zinc dust and stir the solution thoroughly. If enough zinc has been added, the blue color will disappear. The zinc

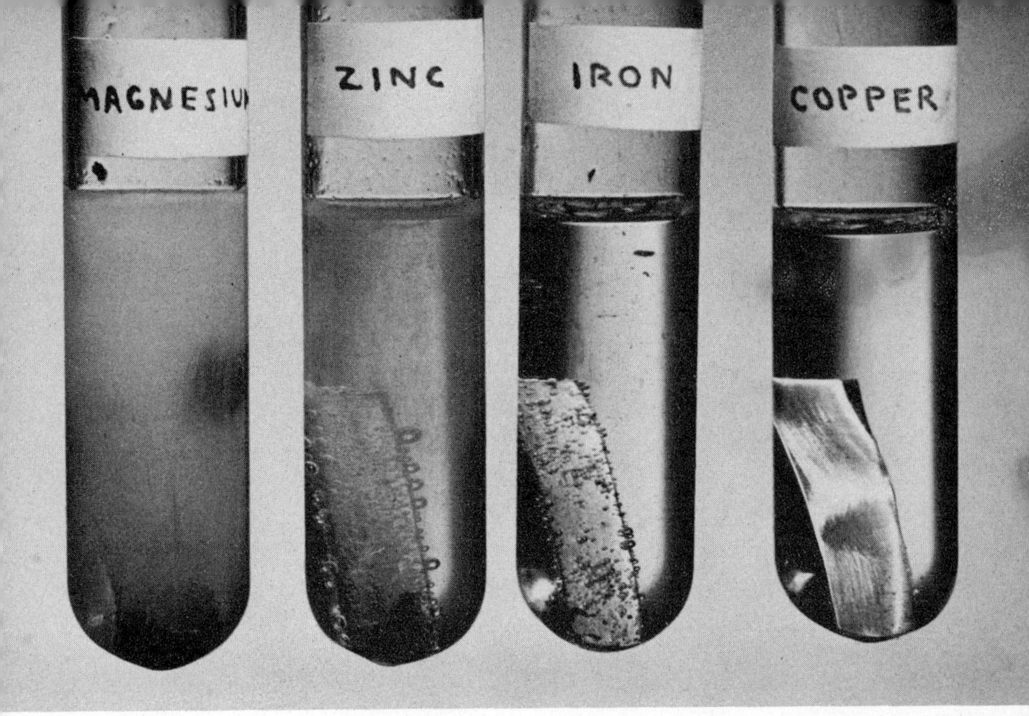


Relative activity of common metals is shown by reaction to acid.

ousts the copper from the copper sulfate and joins with the sulfate radical to form zinc sulfate—which, in solution, is colorless. The ousted copper settles to the bottom of the beaker as finely divided red metal.

**Immersion plating** of one metal on another is accomplished by this changing of places of metals. Iron, for example, can be given a thin plating of copper by immersing it in a solution of 7 g of copper sulfate and 3.5 ml of sulfuric acid in 500 ml of water. A little iron changes places with some of the copper in the copper sulfate, and the displaced copper adheres to the iron.

**Aluminum pots and pans** in the kitchen are often darkened by the swapping of metals. If you cook oatmeal, spinach, or other iron-containing food in an aluminum pot, some of the aluminum changes place with some of the iron, the latter being deposited as a dark coating inside the pot. Although the tidy housewife may scour away this deposit, she needn't do so, for the iron will be removed chemically if some acid food such as tomatoes, rhubarb, or sauerkraut is later cooked in the pot. The iron thus regained is not only harmless but is also a valuable food mineral.

Zinc dust stirred in a blue copper sulfate solution ousts the copper and takes its place. The solution turns water-white.

**Metals give up electrons differently.** A more specific way to express the activity of metals is to say that they vary in their ability to ionize, or give up electrons. Metals at the top of the list give up electrons more easily than those at the bottom. This difference makes electric batteries possible and explains corrosion and electrolytic action between touching metals.

If two metals of different activity are immersed in a suitable solution and then connected by a wire, electrons will flow through the wire from the most active metal. The farther apart the metals in the series, the greater will be the electromotive force or voltage.

**Electrical activity explains corrosion.** A similar electrical effect accounts for the accelerated corrosion that often takes place when two metals are in contact in the presence of moisture. An electric current is set up, causing the more active metal to dissolve more rapidly than normal, while at the same time the less active metal is preserved from corrosion.

You can demonstrate this vividly. Wind a short length of iron wire tightly around a small strip of magnesium (a small bundle of magnesium ribbon will do) and a similar wire around the handle of a silver spoon. Immerse these metal combinations in separate glasses containing dilute sulfuric acid (2 drops to 100 ml of water) with enough potassium ferricyanide to color the solution pale yellow.

If silver is covered with an electrolyte in an aluminum pan, electrolysis will clean off tarnish without removing silver.

Quickly, the solution in the glass containing the silver-iron begins to turn blue, indicating that iron is dissolving and uniting with the ferricyanide. The other solution turns only faintly blue, if at all, for the magnesium, being more active than the iron, dissolves in place of it. This explains why the iron in "tin cans" corrodes more rapidly when the tin plating is broken than if it were not plated at all. It also explains why the iron in zinc-plated, or "galvanized," iron is protected when the zinc coating is damaged. In the latter case, the zinc dissolves, and in doing so forms a protective coating over the iron.

**Clean silver electrolytically.** Solutions and "magic" plates of aluminum or magnesium that are sold to clean silverware without scouring depend upon an exchange of metals. Ordinary baking soda (sodium bicarbonate) and salt (sodium chloride), plus any aluminum pan, will enable you to work the same magic. Place the silver so that each piece touches the pan. Cover with a hot solution of 1 teaspoonful of soda and another of salt to each quart of water. After several minutes, remove the silver, rinse, and polish with a soft cloth.

Touching the aluminum and surrounded by the electrolyte, the silver forms one plate of an electric cell. By action of this cell, the tarnish of silver sulfide is dissolved. Then the sulfur is separated and the silver is redeposited. The method should not be used on oxidized or "French finish" silver, for it may alter the finish as well as clean the surface.

You can easily make and collect oxygen for home lab experiments by means of the generator and pneumatic trough shown above.

# Oxygen Is Everywhere

DO you know that iron and steel will burn; that the rusting of iron produces as much heat as the actual burning of the metal; that the reaction of cold chemicals may produce fire? These are just a few of the facts that can be demonstrated by experiments with oxygen, the most widely distributed and abundant of all the chemical elements.

Where can you find oxygen? It might be easier to list the places and substances where you cannot! Your own body is made up of 65 per cent of this gas by weight. The waters of the sea and the water you drink are nearly 89 per cent oxygen. Oxygen constitutes about 21 per cent of the air you breathe. Oxygen in the earth's crust weighs nearly as much as all the other elements put together—including iron, lead, copper, and all the metallic elements.

Despite its abundance, oxygen was not discovered until a year before the Revolutionary War. Recall that oxygen is a gas without color, taste, or odor, and you can understand how it kept its identity a secret for so long.

**Priestley, an English clergyman, made the discovery.** He placed some red mercuric oxide on a column of mercury in a glass tube similar to a barometer tube and heated the oxide by focusing the sun's rays on it. The column of mercury was pushed down, showing that a gas had apparently been produced. Priestley thrust a glowing splinter into the gas and found that the splinter burst into flames. A mouse placed in the gas suffered no ill effects. Breathing some of the gas himself, the experimenter felt invigorated. A new gas had been found which the famous French chemist Lavoisier soon called "oxygen."

**Oxygen may be made easily** in your own lab by heating a mixture of potassium chlorate or potassium perchlorate with powdered manganese dioxide. Perchlorate, if obtainable, is preferable to ordinary chlorate, as it is a more stable chemical and its release of oxygen is more easily controlled.

Mix together—without grinding—a few grams of the potassium chlorate or perchlorate with about a third as much manganese dioxide. *(Caution: Never grind potassium chlorate or perchlorate with any other substance. The friction may cause an explosion.)* Place this in a Pyrex test tube and clamp the tube almost horizontally on a support, as shown in the photo on the opposite page.

As oxygen dissolves only slightly in water, the gas may be easily collected in a pneumatic trough—a pan of water having a perforated shelf on which is placed an inverted jar of water. Bent glass tubing, connected by rubber tubing to an outlet tube fitted to the gas generator, leads the oxygen under the water in the trough up through the hole in the shelf and into the mouth of the inverted collecting bottle. Hold the water in the jar by means of a sheet of glass or cardboard until the mouth of the jar is under the surface of the water in the pan. As oxygen is generated, it will slowly bubble up into the bottle, gradually replacing the water in the latter.

**To generate oxygen,** heat the test tube gently with an alcohol lamp or bunsen burner turned low, moving the flame along the tube to heat it evenly. Then increase the heating until gas commences to bubble from

In the top photo, a small piece of steel wool that has been ignited glows at red heat in air. Place it in a jar of oxygen, however, as in the lower photo, and it at once bursts into bright flame.

the delivery tube. The first bubbles should be allowed to escape, as these are of air being driven from the tube. The following ones should be directed to bubble up in the collecting jar. The flow of oxygen can be regulated by controlling the heat.

When the jar is full, slip the cardboard or glass plate under it again and turn the jar upright. To prevent the oxygen from diffusing into the air, keep the plate in place until you are ready to use the gas. Before withdrawing the flame from the test tube, be sure to remove the stopper from the tube; otherwise water will be sucked up from the trough and will crack the hot glass.

**Steel burns in oxygen.** Many substances which burn feebly or not at all in ordinary air will burn brilliantly in the oxygen you have collected. Thrust a glowing splinter of wood into it, and the splinter will burst into bright white flame with almost explosive suddenness. Twist a short length of iron wire around a small wad of fine steel wool and ignite the steel wool in your bunsen flame. It will merely glow red. Thrust it quickly into your oxygen tank, however, and it bursts into brilliant flame.

**All oxidation produces heat.** When oxygen combines with other substances so rapidly that the heat of this combination produces flame, we call the reaction "combustion." But *all* oxidation produces heat, whether the heat is apparent or not. The rusting of iron or steel, for instance, produces as much heat as the burning of the same metal. This form of oxidation is so slow, however, that the heat is dissipated as fast as it is generated.

A simple experiment will prove that rusting produces a rise in temperature. Wrap the bulb of a thermometer with fine steel wool; then moisten the wad of wool with 10 per cent acetic acid, which will act as a catalyst to speed up the rusting of the metal. Squeeze out the excess acid, and hang the thermometer in a place free from drafts. Rust will quickly appear, and as it does, the mercury in the thermometer will slowly rise.

**Sodium peroxide starts a fire.** At high temperatures, oxygen is so active chemically that it will unite readily with almost every other element. At ordinary temperatures, it is only moderately active. There are a few oxygen-containing compounds, however, that may react violently with certain other substances even at relatively low temperatures. Sodium peroxide is one. *(Caution: Keep sodium peroxide dry, and away from acids and all kinds of organic substances. If you should spill any, flush it away with a large volume of cold water.)* With a spoon or metal spatula, place a few grains of sodium peroxide on a small wad of absorbent cotton in a can cover. Add a drop or two of warm water. Almost instantly the cotton will burst into flames.

**Place a little sodium peroxide on a wad of cotton in a can cover, put several drops of warm water on it—and stand back! The wet peroxide oxidizes the cotton so fast that it flames up instantly.**

**Small quantities of oxygen** can readily be produced from the ordinary hydrogen peroxide of your medicine cabinet. This chemical, $H_2O_2$, is really water with an additional atom of oxygen. Heat, sunlight, alkalis, and even dirt will cause it to decompose. Powdered manganese dioxide, dropped into a little of the solution in a test tube, will cause oxygen to come off rapidly. The usual 3 per cent solution contains 10 times its own volume of oxygen.

You can prove visually that water is made up of 2 parts hydrogen and 1 part oxygen by means of the simple apparatus shown above.

# Hydrogen—First of the Elements

WHEN hydrogen gas is burned in air, the chemical reaction that produces *fire* also produces *water!* For the ordinary water we drink is nothing but an oxide of hydrogen—$H_2O$, the commonest formula in the catalog of chemistry. The name "hydrogen" itself means "water producer."

Although seldom found free, hydrogen is an important and abundant element. It constitutes one-ninth of the weight of water and two-thirds of its volume, and it is present in all acids and bases. Compounded with carbon, it forms a vital element in all animal life and plant life and in most substances produced by them. It is present in all natural-gas and petroleum products. Flames of incandescent hydrogen a third of a million miles high have been known to burst from the sun's chromosphere.

Pure hydrogen is invisible and odorless, and is by far the lightest of all elements. Its lightness makes it the most efficient gas for dirigibles and balloons. Because hydrogen is dangerously flammable, however, the next lightest gas, inert helium, is used instead where it is available. Mixed in a blowtorch with oxygen, hydrogen makes one of the hottest flames known. Bubbled through liquid oils in the presence of a catalyst, hydrogen changes them into solid fats employed in cooking and soap-making and for lubricants.

**Hydrogen from water.** That water contains 2 parts hydrogen to 1 part oxygen can be shown by decomposing water in apparatus that can be set up in a few minutes. Fill two ⅝- by 6-in. test tubes completely with water that has been made conductive by dissolving in it 1 part of sodium carbonate (washing soda) to each 20 parts of water. Holding the water in carefully with a forefinger, invert them in a glass or plastic container half full of the same solution. Then connect three or four dry cells in series to supply current to two electrodes arranged so one projects upward into each tube. These electrodes may be ordinary 1½-in. nails. Use rubber- or plastic-covered connecting wires, joining them to the nails by wrapping about one inch of a bared end around the head end of each nail and then coating the exposed copper with melted paraffin.

As soon as the battery has been connected, gas bubbles will pour upward from the electrodes, displacing water in the test tubes. Those from the negative electrode will be tiny hydrogen bubbles rising rapidly, while those from the positive side will be large bubbles of oxygen produced less frequently. It will be easy to see that the hydrogen formed is twice the volume of the oxygen.

**Hydrogen from an acid.** An easier way to produce quantities of hydrogen in the laboratory is to displace it from an acid by means of a metal. Dilute hydrochloric or sulfuric acid and zinc are generally employed. Use mossy zinc (irregular lumps made by pouring molten zinc into water) from your lab supply house or cleaned zinc scraps from an old dry cell casing, and cover a small handful of the metal with water in the gas generating bottle shown on the next page. Make this from an 8-oz wide-mouth bottle fitted with a 2-hole stopper. A thistle tube, adjusted to reach to within ¼ in. of the bottom of the bottle, goes through one hole, and a short outlet tube having a right-angle bend through the other.

A few drops of a strong solution of copper sulfate may be added to

Hydrogen is made by reacting zinc with acid in the generating bottle (left) and is then collected in a pneumatic trough.

aid the reaction. By means of a short rubber tube, connect the delivery tube with the bent tube that leads under the water in your pneumatic trough, making sure that the joints are tight.

Then generate hydrogen by pouring concentrated hydrochloric acid carefully through the thistle tube until a vigorous reaction with the zinc begins. *(Caution: If more acid is added later, be sure no air bubbles are carried down the thistle tube, or, better still, don't add more acid!)*

To catch the hydrogen gas, fill several test tubes or jars with water and invert each in turn over the water trough, for hydrogen is collected by downward displacement. Let the gas bubble into the tubes until the water is completely displaced. As each is filled with gas, lift it from the trough still inverted, slip a glass plate over the mouth, and leave it inverted.

**Hydrogen extinguishes flame.** Although hydrogen burns readily in air, it will not support combustion, as can be shown dramatically with a jar of the gas and a match. Support the jar upside down on a clamping stand or with tongs to keep from burning your hands, and introduce into the opening a lighted match held in the end of a glass tube. The gas will light at the mouth immediately, but the match itself will be extinguished if pushed well into the jar. Withdrawn again, the match will be relighted by the flaming gas at the mouth.

**Water made from fire.** Hydrogen is highly explosive when mixed with air, and for that reason it is imperative that it be handled with care. Read

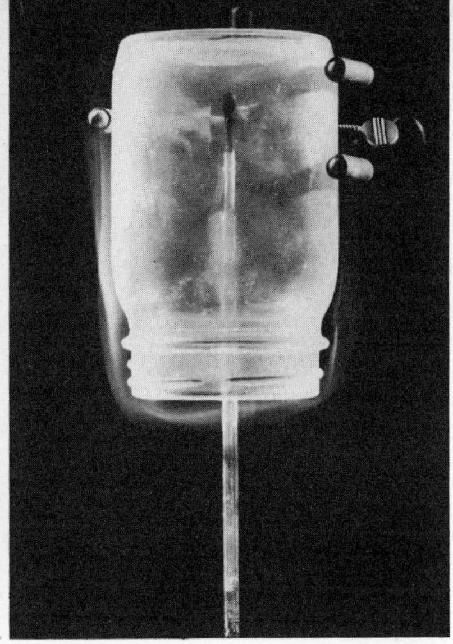

Bring a lighted match up to an inverted jar of hydrogen, and the gas ignites; but push the match inside and its flame goes out.

carefully and always follow the precautions displayed on page 51. After that you can burn hydrogen in air to prove that it produces hydrogen oxide or, in common words, plain water. Connect a drying tube filled loosely with granular calcium chloride to the delivery tube of your hydrogen generator, and to the other end of the drying tube connect a bent glass tube drawn out on its free end to a jet.

Never attempt to light the gas from the jet until you have tested it to make sure the hydrogen is unmixed with air. To do this, connect the jet to the tube from your pneumatic trough by means of a short rubber tube and collect a test tube of gas as before. Keeping the tube inverted, remove it several feet from the generator and light the gas at the mouth of the tube. If the gas explodes even slightly, it still contains air. If it burns quietly, use this flame as a torch to light the jet; *never use a match.* As a further safety measure, wrap the generator loosely with a towel to catch the pieces if it should break.

*After the gas has been proved pure,* light the jet. If a metal or ceramic jet is used, the flame will be an almost invisible blue. A glass jet may impart a yellowish tint because of the sodium in the glass. Now hold a cold tumbler or beaker over the flame, and drops will form on the inside of the glass. These drops can be shown by tests to be pure water—water made from fire! To prevent a flashback, finally extinguish the flame with a bit of wet cloth or paper.

**Hydrogen makes balloons rise.** If you would like to show the remarkable lifting power of hydrogen and inflate toy balloons with it, a special

**To a chemist, water is hydrogen oxide. To prove that water can be made by burning hydrogen, hold a cold tumbler or beaker over a hydrogen flame. Drops of water will condense on the inside.**

generator for the purpose can be made by fitting an ordinary 8-oz bottle with a 1-hole stopper into which is pushed a drying tube containing calcium chloride. Put a few pieces of mossy zinc in the bottle, cover them with dilute hydrochloric acid (about 1 part acid to 4 parts water), and add a few drops of copper sulfate solution.

Tie a balloon to the open end of the drying tube and hold the stopper to keep it from popping out while the balloon is being inflated. When you remove the balloon, tie its neck to prevent the gas from escaping. A small balloon will lift itself and at least several paper clips. The approximate lifting power of the gas is about 1.2 g per liter, since hydrogen weighs 0.09 g per liter and air 1.29 g.

**Small toy balloons can be inflated with hydrogen by means of this generator provided with a drying tube.**

## DANGER—HIGH EXPLOSIVE!

• Mixed with air, hydrogen is dangerously explosive, but it is an interesting and important gas made in all school and most home laboratories, and it is absolutely safe when handled properly. Accidental explosions are caused by either ignorance or carelessness. If you learn the following simple rules and observe them faithfully, you will never have cause to worry.

• Don't generate large quantities of the gas. An 8-oz generator should be the largest used in a home lab.

• Make all connections gastight.

• Never ignite hydrogen issuing from a generator until a sample collected in a test tube has been burned with a quiet flame.

• At other times, never allow a flame near the generator.

• It is best not to add a second charge of acid to a generator. If you must, be sure no air bubbles are carried through with the acid.

• To prevent a flashback, extinguish a hydrogen flame with a piece of wet paper or a wet cloth.

• Don't try to produce hydrogen-air explosions in anything but a test tube or other small straight-sided vessel.

• Wrap the generator loosely with a towel to catch glass if it breaks. This should never happen, however, if all the other rules are obeyed.

Chlorine is generated in the flask and flows into the center jar when the pinchcock is open. Waste goes into the scrubbing tube.

# Chlorine Takes Life and Saves It

LIFESAVER and killer—that is the Jekyll-Hyde role of chlorine, the most plentiful and useful of the halogens. In peacetime industry, more than 1,500,000 tons of chlorine are used yearly for bleaching paper and cloth, in making dyes, solvents, insecticides, fire-extinguishing liquids, and other organic chemicals, and in purifying water. In warfare—should it become necessary to use deadly gas—phosgene, mustard gas, and lewisite are compounds of chlorine. Titanium tetrachloride produces dense white smoke screens.

**Chlorine gas is produced commercially** almost entirely by electrolysis (see page 73). In one process an electric current is passed through a strong solution of common salt—producing chlorine at the anode,

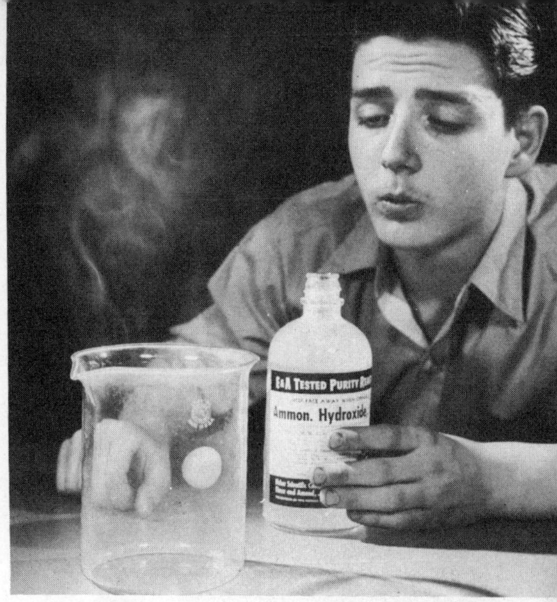


Chlorine unites with hydrogen from turpentine, producing flame. Hydrogen chloride thus formed makes white cloud with ammonia.

hydrogen at the cathode, and changing what remains of the salt to sodium hydroxide. In another process a current is passed through pure molten salt, so breaking the salt into its two constituents, sodium and chlorine.

**Chlorine may be produced more easily** in your laboratory, however, by heating a mixture of hydrochloric acid and manganese dioxide. With apparatus set up as shown, you may obtain gas as you need it and prevent excess gas from escaping into the room. Although chlorine is safe if handled properly, remember it is a poison and may cause serious throat and lung irritation if you breath much of it. All the joints of the apparatus must be gastight, and chlorine not actually used should be led out through a window or absorbed in a scrubbing column.

Put about 10 g of granular manganese dioxide in a flask that has a stopper fitted with a bent glass delivery tube and a thistle tube which extends nearly to the bottom of the flask. The scrubber tube, at the right of the apparatus, should be about 1 in. in diameter and 12 to 18 in. long. It is filled loosely with lump lye or lumps of moist lime. A pinchcock in the setup allows chlorine to pass into either the scrubber or a container for your experiments. To generate chlorine, pour 25 ml of concentrated hydrochloric acid into the thistle tube, swirl the flask a little to mix the chemicals, and heat the flask gently. Place the central outlet tube in a glass jar, passing it through a hole in a loosely fitted cardboard cover. A light behind the jar will show the heavy, greenish-yellow gas collecting.

Chloroform is made by gently distilling a mixture of chloride of lime and acetone. It can be recognized by its sweetish smell.

**Chlorine produces flame.** Chlorine is such an active element that some metals, if powdered, combine with it spontaneously, flashing into flame and leaving a residue consisting of the chloride of the metal. A little "gold" bronzing powder will do this if dropped into a jar of the gas. Copper, zinc, and iron will react with chlorine if heated slightly. Thrust a heated copper wire into a jar of chlorine and the wire will immediately become coated with green copper chloride.

**Chlorine has a particular affinity for hydrogen.** To prove this, attach a little wad of cotton to a wire handle, soak the cotton in quite warm turpentine (heat the turpentine by immersing a small beaker containing a few milliliters of it in a dish of boiling water), and then dip the wad into a jar of chlorine. Hydrogen will be drawn so violently from the turpentine that the wad will burst into smoky flame.

To demonstrate that the chlorine has combined with hydrogen to form hydrogen chloride, blow the vapor from an unstoppered bottle of ammonia over the glass in which the turpentine has burned. The dense white cloud of ammonium chloride that pours from the glass is proof.

**Chlorine makes chloroform.** In some reactions chlorine can not only pull hydrogen from compounds but can displace it with itself. For instance, chlorine can displace three of the four hydrogen atoms in the gas methane ($CH_4$) to form chloroform ($CHCl_3$), or all four of the hydrogen atoms to form carbon tetrachloride ($CCl_4$). If you have a condenser and running water, you can easily demonstrate this type of displacement by making a sample of chloroform.

In a 250-ml flask put a mixture of 20 g of bleaching powder (household "chloride of lime") and 30 ml of water. Add 4 g of acetone. Then mount the flask on a ring stand and connect a condenser, as shown. Place a small test tube or beaker at the outlet of the condenser to catch the distillate.

Heat the flask cautiously with a *small* flame until the mixture begins to froth, remove the flame until the reaction moderates, then heat the flask again until its contents boil. Continue boiling until no more oily drops come from the condenser. Two layers of liquid will be found in the receiving test tube, the lower one being the chloroform. Pour some out and you will recognize it by its characteristic sweet smell.

**Chlorine as a bleach.** Large quantities of chlorine are used for bleaching paper, viscose rayon, and cotton. To show how chlorine is used to bleach natural cotton—the chemical tends to destroy silk and wool—set up four tumblers in a row. In the first put a 10 per cent solution of bleaching powder; in the second a 5 per cent solution of sulfuric acid; in the third a 5 per cent solution of sodium sulfite; in the last put plain water.

Now immerse a strip of unbleached muslin in the first solution until it is thoroughly wet. Notice that the cloth is not bleached. Transfer it to the second glass, however, and the cloth rapidly lightens. This is because



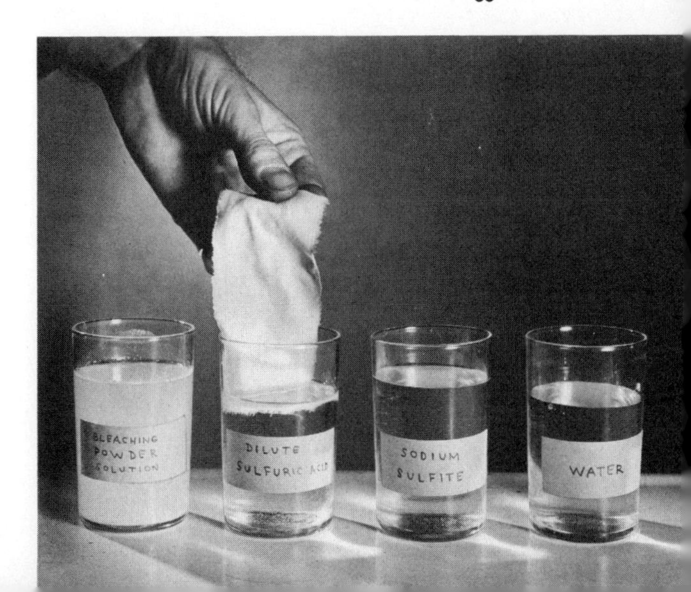


**Unbleached cotton, immersed in the bleaching powder solution at left, is bleached by action of sulfuric acid in the next glass. Remaining chemicals are removed by sodium sulfite and plain water.**

Chlorine will not bleach unless water is present as a catalyst. At left, two identical strips of dyed cotton—the one at right in the photograph dry, the other moist—are suspended in tumblers of the gas. The dry cloth keeps its color, the other is bleached.

the sulfuric acid sets free the chlorine in the bleaching powder, forming a solution of hypochlorous and hydrochloric acid. It is the hypochlorous acid that does the bleaching by oxidizing the natural coloring matter in the cotton, thus changing it into a colorless compound. Immersion of the cloth in the sodium sulfite solution will remove the excess chlorine. Final rinsing in plain water will remove the other chemicals.

To test the bleaching power of chlorine on different dyes, suspend strips of moistened dyed fabrics in jars of the gas. Although water takes no active part in the bleaching, its presence is always necessary. Dyed material that will bleach out completely in chlorine when moist, will not bleach at all when dry. The moisture acts as a catalyst.

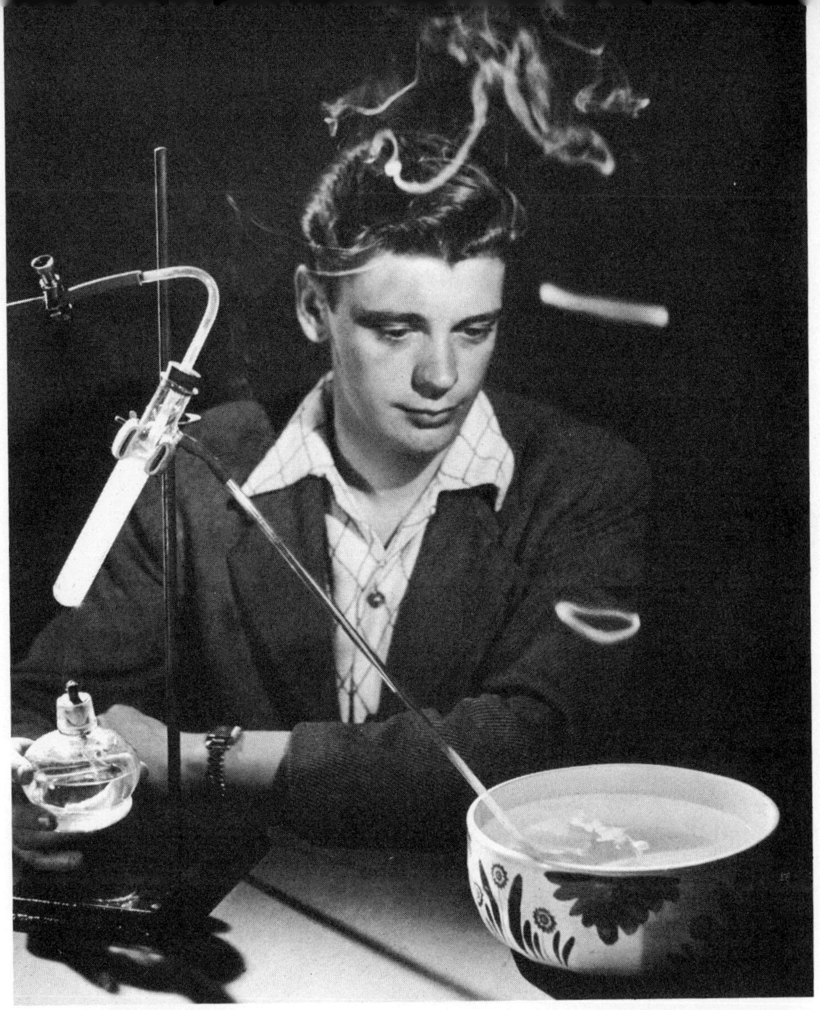

Phosphorus heated in a solution of lye produces phosphine, which ignites spontaneously on contact with air, forming smoke rings.

# Phosphorus—the Alchemist's Element

PHOSPHORUS, the strange waxlike element that glows weirdly in the dark and ignites spontaneously, accompanied by clouds of dense white smoke, was discovered, appropriately enough, in an alchemist's musty laboratory nearly three centuries ago. Brand, of Hamburg, chanced upon it while experimenting in search of the mythical philosophers' stone.

Today, phosphorus is found abundantly in compounds. Too active, chemically, to be found in the uncombined state, the element exists in all animal and vegetable life, in earth deposits as phosphate rock, in most

Phosphorus in the bottle cap floating on warm water catches fire, producing a dense "smoke" of phosphorus pentoxide. If you cover the cap with the tumbler, the smoke will dissolve in the water to form phosphoric acid.

river and spring water, and in the sea. Bones are made up of about 58 per cent calcium phosphate. In Florida, South Carolina, and Tennessee there are great deposits of phosphate rock, probably the fossil remains of prehistoric animals.

**Handle phosphorus with care.** An ounce of white phosphorus in thin sticks is sufficient for dozens of experiments. As the chemical is extremely poisonous and combustible, these precautions must always be observed: Take care to store it under water; *always* handle it with forceps or tongs, *never* with the fingers (phosphorus burns are poisonous and difficult to heal); cut it only under water; return unused pieces to the water-filled bottle immediately; and burn waste bits.

**How phosphorus produces smoke** for screening purposes in war, and how this smoke may be dissolved in water to form phosphoric acid for peacetime uses, can easily be shown. With tongs, drop a pea-size piece into a metal bottle cap (from which the liner has been removed) floating in a dish of warm water. The phosphorus will catch fire immediately. If you now cover it with a large inverted tumbler, dense white clouds of phosphorus pentoxide will quickly fill the glass.

This "smoke," which is really a light, finely divided powder, combines avidly with water, drawing some of the water from the dish violently into the tube, and will disappear in solution after several minutes. Testing this water with litmus paper will disclose that it has become strongly acid. The acid is phosphoric acid, which is commercially prepared in its purest form by dissolving phosphorus pentoxide in distilled water. Because of its great affinity for water, phosphorus pentoxide powder is often used for drying gases.

**Phosphorescent paint.** If exposed to air at temperatures above about 30°C, white phosphorus catches fire spontaneously. Below this temperature, slow oxidation causes its surface to glow in the dark. This glow, or phosphorescence, may be demonstrated more vividly with a "magic" paint. In a test tube dissolve a piece of phosphorus, half the size of a pea, in 5 ml of carbon disulfide (*keep far away from flame*). Add 5 ml of olive oil, and shake gently to mix. With a small soft brush, paint with the mixture on heavy cardboard, being sure to immerse the brush in water immediately after use to prevent spontaneous combustion, and clean it later in carbon disulfide and then in alcohol. In the dark, the letters or figures on the cardboard will glow with a weird, pale light. Blow on them, and they disappear—only to reappear again when you stop blowing.

**Fighting a phosphorus fire.** Though quick to ignite, phosphorus has a relatively low heat of combustion and most of its heat is directed upward. It therefore is not a good incendiary substance except when in contact with flimsy materials such as paper, cloth, grainfields, and small pieces of wood. The best way to fight a phosphorus fire is to protect nearby com-

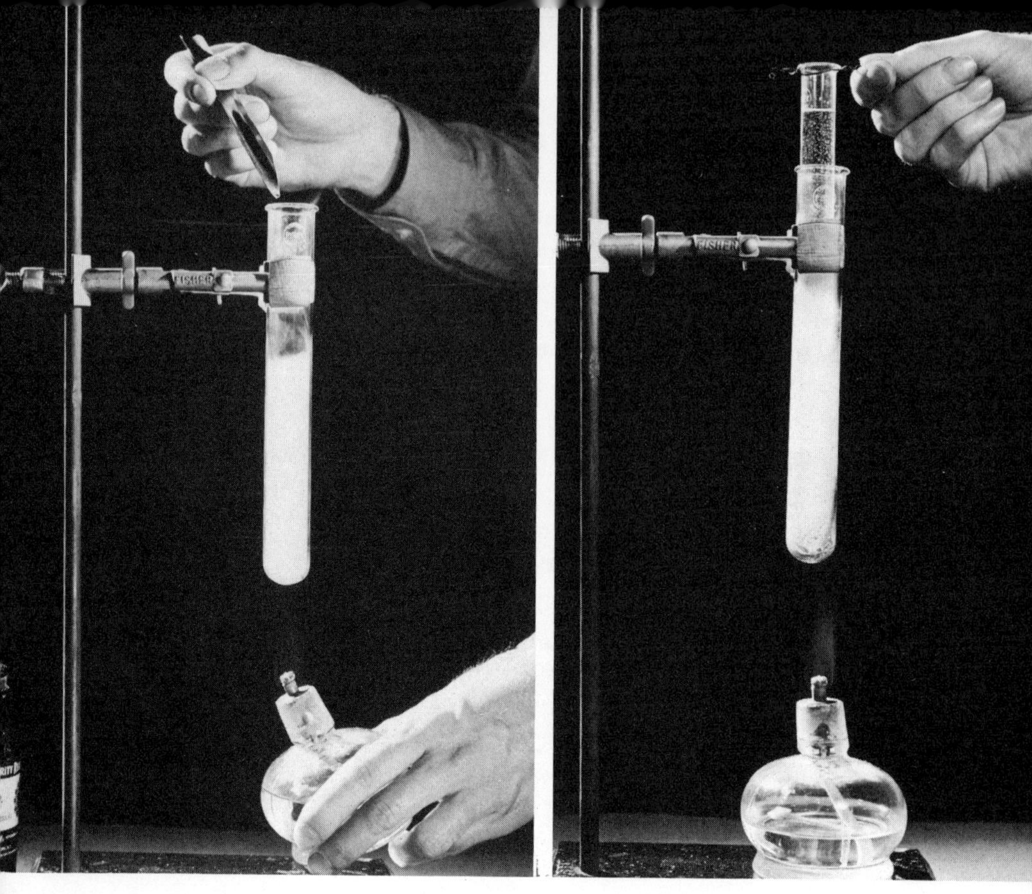

Adding iodine helps change white phosphorus into the red variety. At right, continued heating changes the red phosphorus back to white, which condenses on the smaller tube inside the other.

bustible material with a water spray while allowing the phosphorus to burn itself out in its own time.

**Red phosphorus is different.** Strangely enough, poisonous and readily combustible white phosphorus has a brother, red phosphorus, that is not poisonous and does not catch fire until heated to about 330°C. Red phosphorus is now used in great quantities in matchmaking, replacing white phosphorus, which was a source of widespread bone disease among match workers.

Mount a large test tube vertically on a stand, drop in two pea-size pieces of white phosphorus, and apply a small flame. The phosphorus may start to burn, but its own inert fumes will soon cut off oxygen. Under continued heating, the white phosphorus will gradually turn reddish. Add a speck of iodine crystal, as a catalyst, and the change from white to red phosphorus will be almost instantaneous.

To change the red phosphorus back to white, fill a smaller test tube with cold water, twist a piece of wire around its neck for a support, and lower it into the large tube. Heat the large tube more strongly than before, and fumes will arise from the red phosphorus and condense as crystals of white phosphorus on the outside of the cool inner tube. When the whole apparatus has cooled, lift out the small tube by means of tongs, and the white phosphorus on its surface will ignite spontaneously a few seconds after contact with the air.

**Phosphine makes fire and smoke rings.** Phosphine, a gaseous compound made by uniting phosphorus with hydrogen, may be the subject of a spectacular experiment, which must be performed in a well-ventilated room and with care paid to details.

Half-fill a side-necked test tube with a strong solution of lye or sodium hydroxide, allow it to cool, and add a pea-size piece of phosphorus. Fit the tube with a 1-hole stopper through which passes a glass tube reaching nearly to the bottom of the test tube. Clamp the test tube to a ring stand, as shown in the photo, and connect its side arm to a long glass delivery tube having an upward bend at its lower end. This end of the delivery tube is placed in a pan of water with its opening submerged about a quarter of an inch.

Connect the tube from the test tube stopper to a source of illuminating gas through a rubber tube having a screw pinch clamp. Make sure all connections are absolutely tight; then flush all air from the system by allowing illuminating gas to bubble gently through it for a minute or two, after which turn off the gas at the source of supply and close the pinch clamp tight. This flushing prevents combustion of the phosphine.

Now heat the test tube gently. Bubbles of gas will soon rise in the solution and pass through the delivery tube and up through the water in the pan. After a few minutes of operation, each bubble escaping through the water will burst into a little puff of flame as it comes in contact with air, the flame often changing into a beautiful smoke ring.

Pure phosphine itself does not ignite spontaneously at room temperature; it is a trace of a related compound that sets it afire. Although most of the phosphine generated is burned in this experiment, the gas is poisonous and therefore the experiment should not be continued too long. To stop the reaction, carefully fill the dish with water and remove the flame from under the test tube. When the vapors cool, water will be drawn into the apparatus, cooling it and nearly filling it. The stopper may be removed when the apparatus is thoroughly cold.

You can make a sample of glass by melting together equal parts of sand, sodium carbonate, and lead oxide in a porcelain crucible.

# Silicon Is Sociable

ALTHOUGH silicon can never be found alone, its compounds make up more than a quarter of the earth's crust. Common sand is silicon dioxide, or silica, as are the minerals quartz, agate, amethyst, and opal. Almost all rocks, except limestone, contain this vital element. Sandstone is merely silica bonded with clay or lime. Feldspar and clay are both compounds of silicon and aluminum. Carborundum, used as an abrasive, is an artificial compound produced by heating together silicon dioxide and carbon in an electric furnace.

One of the most important and familiar artificial compounds of silicon is glass. Ordinary window and bottle glass is a mixture of silicates produced when white sand (which is almost pure silica) is melted with sodium carbonate and lime. When potassium oxide and lead oxide are

substituted for sodium and lime, glass with a low melting point and high index of refraction is formed, which is especially suitable for lenses. Different types of glass may be made by other slight alterations in the composition.

**How to make glass.** With a small porcelain crucible and a Fisher burner, you can easily make and color bits of glass in your home laboratory just as it is done in industry.

Because of its lower melting point, it is better to experiment in making lead glass, rather than glass containing lime. Mix about 3 g of clean white sand with equal weights of dry (anhydrous) sodium carbonate and yellow lead oxide. Put this mixture in a crucible and heat it strongly until it melts into a fluid mass. Pour the contents on the bottom of a pie tin or on an asbestos mat, and the drops that form will become, upon cooling, sparkling bits of real glass.

**You may color your glass** by adding to it a trace of some metallic oxide while it is still in its molten form. Cobalt oxide produces blue glass; chromium oxide colors it green. Used in minute quantities, manganese dioxide produces amethyst glass, while larger quantities color it black. Colloidal gold turns glass a beautiful ruby red. The bright red of traffic lights and automobile taillights is often produced by the addition of selenium.

**Compounds of silicon** owe much of their usefulness to the fact that chemically they are extremely stable. They resist all common acids, except hydrofluoric, and they are also little affected by bases, except hot sodium or potassium hydroxide. This inactivity explains why most chemical containers and laboratory apparatus are made of glass; and why

If you pour together a solution of dilute hydrochloric acid and a solution of sodium silicate, a solid white mass forms almost immediately that will support the stirring rod, as shown. This is silica gel, a substance which, in its dry form, is used as a dehydrating agent.

buildings of glass, brick, and stone have been able to weather the centuries.

The resistance of silica to acids and bases makes it of special value to the carpenter and home craftsman. In the form of silex, an extremely fine sand made from quartz rock, it serves as a base for wood fillers that is totally unaffected by vapors in the atmosphere or chemicals in paint.

**Sodium silicate makes a cement.** Of all the silicates, only potassium and sodium silicates are soluble in water. Sodium silicate solution is the well-known "water glass," used extensively for making cements, for fireproofing, and for preserving eggs. With the syrupy variety, you can make cement for glass and china. Dilute it, and you can produce a gel and a chemical garden.

You can use thick water glass alone as a good transparent cement for glass or china. First heat the edges to be cemented; then apply the water glass. Clamp the parts tightly together until the water glass is dry. A cement that will withstand acids and high heat may be made by mixing two parts of the thick water glass with one part of fine sand and one part of ground asbestos.

**To make a silica gel,** two solutions must be prepared. The first is made by diluting 15 ml of thick water glass with an equal amount of water. The second consists of 2 ml of concentrated hydrochloric acid diluted with 20 ml of water. Pour these two solutions simultaneously into a small glass or beaker, and immediately stand a stirring rod in the center of the beaker. Within a few seconds, the clear solutions will have united to form a solid whitish gel that will support the rod and will not fall out if the beaker is inverted. When dried, this silica gel contains millions of microscopic pores that will take up quantities of water. It is widely used as a drying agent.

**A chemical garden** affords a beautiful experiment which no home or school chemist should miss. Obtain a small fish bowl or a low jar that will hold about a quart and sprinkle a layer of coarse sand about a quarter of an inch deep on the bottom. Fill the remainder of the bowl or jar almost to the top with water glass diluted with an equal amount of water.

"Seeds" for your garden consist of salts of the heavy metals, such as copper sulfate, cobalt chloride, ferrous sulfate, zinc sulfate, and manganese chloride. For best results, crystals of these chemicals should be at least 3 to 5 mm in diameter. All are of different colors, and should be dropped to the bottom of the globe so that the colors will be distributed attractively.

Salts of metals in sodium silicate solution make a fairy garden.

Within seconds, some of these crystals will be sending up shoots; in minutes, a few may reach the top of the solution; while in an hour or two the garden should be completely grown—a forest of intricate and varicolored growth that suggests an underwater fairyland.

After a day, the sodium silicate solution should be carefully siphoned off and replaced with fresh water. As the "plants" are metallic silicates, they will not dissolve. They should last until they are broken by jarring.

**The process of physical chemistry** that produces this growth is quite complex. When a crystal of copper sulfate, for example, is placed in a solution of sodium silicate, some of the surface dissolves and the crystal is soon surrounded by a concentrated solution of copper sulfate. The copper and the silicate then combine to form a continuous film of copper silicate around the crystal. This acts as a semipermeable membrane through which the outside solution can get in faster than the more concentrated inside solution can get out. Finally, the pressure of the inside solution becomes so great it breaks through the membrane at its weakest point. Once outside, the copper again reacts with the silicate to form a new membrane a little farther from the original crystal. By repeated breakings and mendings, the process continues, until at last the crystalline offshoot reaches the surface of the solution.

Chlorine from the flask displaces bromine from a bromide in the horizontal tube; bromine then displaces iodine from an iodide.

# The Active Halogens

"HALOGENS," or "sea-salt producers," is the name well given to a family of nonmetallic chemical elements that includes fluorine, chlorine, bromine, and iodine. Salts of all these elements do exist in the sea. Sodium chloride, which is table salt, and salts of bromine are taken commercially from sea water, and not so long ago most of our iodine was obtained by heating seaweed ash with sulfuric acid and manganese dioxide. Fluorine compounds, however, are generally found as the minerals fluorspar (calcium fluoride) and cryolite (sodium aluminum fluoride).

**Fluorine is the most active element.** All four elements are so active that none occurs free in nature. Fluorine, a pale, greenish-yellow, corrosive

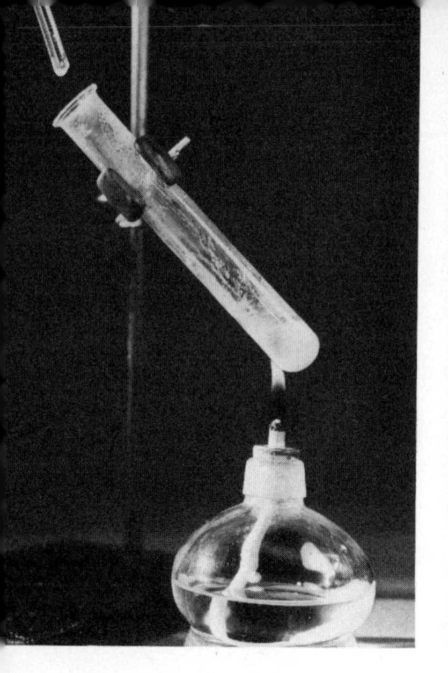

Fumes from fluorides heated with sulfuric acid turn a drop of water white, as at left. Above, apparatus for making bromine.

gas, is indeed the most active of all known elements. This limits its commercial use, but its compounds are valuable. Cryolite is an essential flux in the production of aluminum. Small quantities of fluorides are found in healthy bones and teeth. Hydrofluoric acid is used for etching and frosting glass.

**A test for fluorides.** The ability of hydrofluoric acid to dissolve ordinary sand to form silicon tetrafluoride, or "sand gas," provides the basis for a simple test for metallic fluorides. Put a little of the suspected chemical in a test tube, add a pinch of sand and some concentrated sulfuric acid, and heat the tube gently. If a metallic fluoride is present, the sulfuric acid will combine with it to form hydrofluoric acid. This joins with the sand to form sand gas. A drop of water held at the mouth of the test tube will now turn white, due to the formation of whitish silicic acid.

**Chlorine,** a greenish-yellow gas slightly darker than fluorine, is the next most active halogen. Because of its many important uses it has been treated in a separate unit of this book.

**Bromine is a fuming liquid.** The third member of the family, bromine, is a dark, reddish-brown, fuming liquid. Its name comes from the Greek word *bromos,* meaning "bad smell." The fumes are extremely poisonous —so much so that they are employed as a disinfectant and as a war gas.

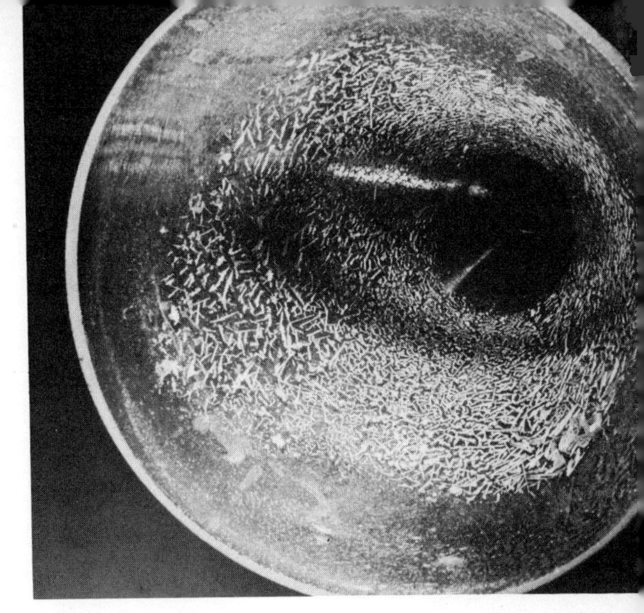

**Iodine is made by heating an iodide with manganese dioxide and sulfuric acid. Iodine crystals collect on the inverted funnel.**

Bromides are used widely in photography and as nerve sedatives. Vast quantities of bromine, derived from sea water, are used to make ethylene dibromide, a vital ingredient in antiknock gasoline.

**To produce bromine,** mix 1 g of potassium of sodium bromide with 1 g of powdered manganese dioxide and put it in a test tube. Add 2 ml of sulfuric acid to 1 ml of water in another test tube and, after the mixture has cooled, add it to the powder.

Now set up your tube on a stand, as shown in the photo on page 68. A bent delivery tube attached to the test tube through a 1-hole stopper is then led into another test tube half filled with water and kept cool in a glass of cold water.

Heat the mixture very gently, being careful not to boil it through the delivery tube. Dark fumes will pour through the apparatus, dissolving in the water in the collecting tube and coloring it a reddish brown. Pure bromine is dangerous to touch or smell, but this diluted bromine is safe to smell cautiously and will not injure the skin if washed off promptly.

**How to make iodine crystals.** Although everyone is familiar with the brown liquid tincture of iodine (a solution of iodine in alcohol), pure iodine, a purplish-black crystalline solid, is seldom seen except by chemists. Make it like bromine, but use a beaker or an evaporating dish with a cold funnel inverted over it to condense the vapor. Use the same proportions of chemicals as you did in making bromine, substituting sodium or potassium iodide for the bromine salt, and heat the mixture with a very small flame. Dense clouds of violet vapor will arise and condense on the cold funnel, leaving after a few minutes a pattern of hundreds of beautiful crystals of iodine. They may be scraped off and used in experiments. Oddly enough, iodine will turn directly from a solid to a vapor before it melts, and then it may be condensed back into a solid by cooling the vapor. The complete process is called "sublimation."

**Chlorine frees bromine and iodine.** Because chlorine has a greater affinity for the metallic portion of a compound than has either bromine or iodine, it will free both of these substances from a metal and join with the metal itself. Bromine, having a greater affinity for metals than iodine, will set iodine free.

You can demonstrate vividly the relative activity of these three elements by means of the setup shown on page 67. Chlorine, generated in the flask at the left, is led through a bent delivery tube into a large horizontal tube. This tube contains, first, several grams of potassium bromide held in place by a loosely packed cotton plug on each side of it and then, several inches farther on, a like portion of potassium iodide similarly held between cotton plugs. A small piece of blotting paper, moistened with a starch solution, is placed in the far end of the horizontal tube to act as an indicator. Then a bent tube leading from the far stopper is connected to an inverted funnel the mouth of which dips just below the surface of a strong sodium hydroxide solution. The latter solution is used to trap excess gas.

To make chlorine, add a little hydrochloric acid through the thistle tube to a water solution of ordinary household chlorinated lime in the flask. As chlorine generated by the reaction passes into the horizontal tube, bromine will be freed from the potassium bromide, and its brown fumes will fill the tube between the two sets of plugs. Driven on by the pressure of the chlorine, the bromine will penetrate the potassium iodide, freeing iodine, the vapor of which will react with the starch paper, turning it black—a familiar test for iodine.

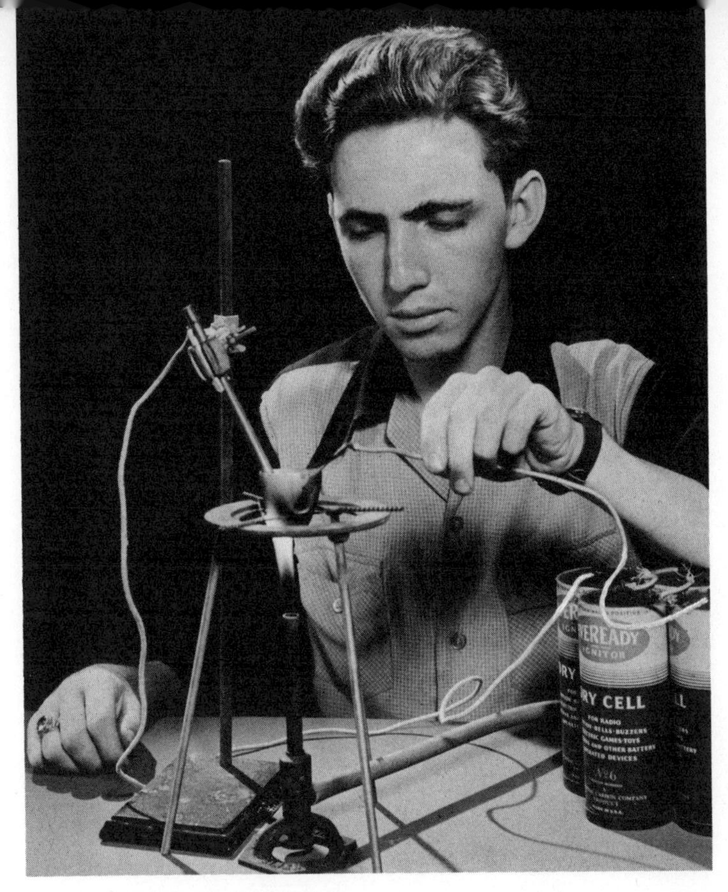

**By means of this simple electrolytic apparatus, you can make small quantities of the element lithium from lithium chloride.**

# Electricity and Chemistry

BY sending an electric current through certain melted salts or solutions of salts in water, the modern chemist performs an almost endless variety of industrial chores. He extracts metals from their ores, plates expensive and hard metals on baser or softer metals, and produces vast quantities of such important elements as aluminum, magnesium, sodium, chlorine, hydrogen, and sodium hydroxide.

**How solution splits compounds.** All these transformations are accomplished by "electrolysis," a process in which electricity changes the chemical composition of a conducting solution through which it passes. The principle of electrolysis is simple and fascinating. When certain salts are fused, and when acids, bases, or salts are dissolved in water, the compounds promptly break up, or "dissociate," into "ions." These ions are atoms or groups of atoms which have become positively or negatively charged because of the loss or gain of electrons. For example, when common salt (sodium chloride) is dissolved in water, the molecules dissociate into sodium ions and chlorine ions. In this process of breaking up, each chlorine atom grabs an extra electron and adds it to the total normally carried when combined with a sodium atom as a salt molecule. This additional electron gives the chlorine ion a single negative charge. Left with a deficit of one electron, the sodium atom consequently attains a single positive charge.

**Because of the electrical charges they carry,** ions behave quite differently, both chemically and electrically, from the neutral molecules. Put electrodes connected with an electric generator or battery into an ionized solution, or "electrolyte," and what happens? Negatively charged ions move toward the positive electrode, or "anode," and positively charged ions toward the negative electrode, or "cathode." Reaching their respective electrodes, the ions are neutralized, and presto! they are promptly changed back into plain atoms.

**Make lithium by electrolysis.** As your first experiment with electrolysis you could make a little lithium. Except for frozen hydrogen, this is the lightest solid known. The apparatus you will need is shown in the photo on page 71. A small porcelain crucible over a bunsen or Fisher burner will serve as your furnace, a carbon rod from a commercial movie projector (obtainable from laboratory and theater supply houses, or perhaps from your local movie) as your anode, and a short length of iron wire with a small loop bent in its lower end as your cathode. Four dry cells connected in series will furnish adequate voltage. The anode should, of course, be connected to the positive terminal of this battery, and the cathode should be connected to the negative terminal.

Melt about 7 g of lithium chloride in the crucible, and then insert the electrodes so that they dip well into the solution but do not touch. Soon after you have immersed the electrodes, a silver substance will begin forming on the loop at the end of the cathode. This is the metal lithium.

**When a little bead has formed,** remove the wire carefully, immerse it in a small bottle containing kerosene, and shake the wire to dislodge the globule of metal. Because lithium reacts with the moisture in the air it should always be kept under kerosene when not being used. Continue collecting the metal until you have enough for a fair-sized bead. Then remove the bead from your bottle with a pair of tweezers and place it on the surface of some water in a beaker. It will float high and dart about, releasing hydrogen and forming lithium hydroxide with the water.

**Chlorine and sodium hydroxide from salt.** By passing an electric current through a strong solution of salt water, chemists produce chlorine, hydrogen, and sodium hydroxide on a large scale. Chlorine forms at the anode while the sodium produced at the cathode reacts with the water, yielding sodium hydroxide and hydrogen. A porous partition of asbestos cloth between the electrodes keeps the chlorine from mixing with the sodium hydroxide.

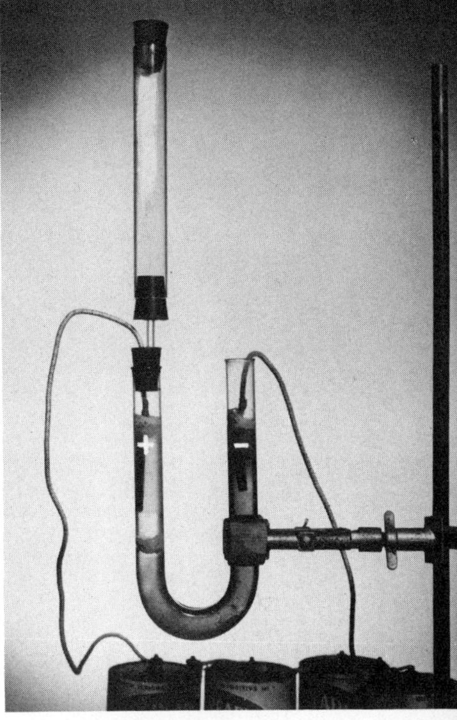


Light lithium floats on water, decomposing it as it does so. Lower photo, current from a battery passed through a salt solution yields chlorine, sodium hydroxide, and hydrogen.

You can readily produce all of these with the help of a U-tube, some strong salt water, and your four-cell battery. Your electrodes can be made from the carbons taken from two flashlight cells. Apply a coating of paraffin to the connecting wires at the points where they are wrapped around the electrodes. This protects the wires from the solution. A disk of asbestos slightly larger than the inside diameter of the tube should be carefully wedged about two-thirds of the way down the anode arm.

**You may test for chlorine** by causing this gas to bleach out a strip of colored cloth in a length of tubing placed above the anode leg of the U-tube and connected to it by a small glass tube. The top stopper in the upper tubing should have a small hole in it to let the chlorine pass out slowly. You may obtain a

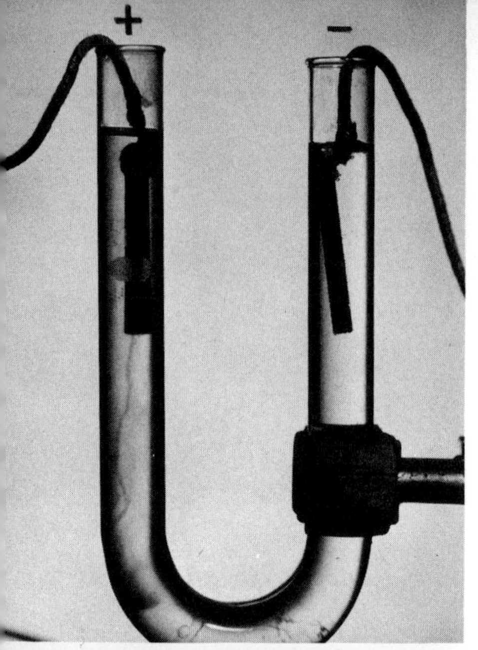

Top photo, electricity produces iodine from potassium iodide. Lower, electricity passed through a solution of tin chloride makes a glittering "tin tree."

more positive bleaching action by dying a piece of cloth with ink and using it while still moist. A few drops of phenolphthalein solution added to the electrolyte in the cathode arm of the tube will give you a test for sodium hydroxide. A strong pink coloration will indicate that sodium hydroxide has been produced.

**Iodine from potassium iodide.** By putting a solution of potassium iodide in the U-tube instead of the salt water you will get a reaction that is plainly visible. As soon as the electrodes are connected to the battery, negative iodide ions are attracted to the positive electrode. When they reach this, they are neutralized, giving up their negative charge and becoming common atoms of iodine. You with recognize this iodine as it streams down.

**A "tin tree" by electrolysis.** Using the same apparatus you can also grow a pretty "tree" of tin within a few minutes. This time use a solution of tin chloride (stannous chloride). You can make this by dissolving about 14 g of the chemical in 125 ml of water, then adding just enough hydrochloric acid to cause the solution to become clear. Use a bare copper wire for the cathode instead of carbon.

As soon as the battery is connected, tin crystals will begin to form on the copper wire. A tree several inches long should form while you watch.

**Plating metals by electricity** is one of the most important and familiar branches of electrolysis. The electrochemical activity in this case is the same as in the previous examples. Electrodes are placed in a solution of

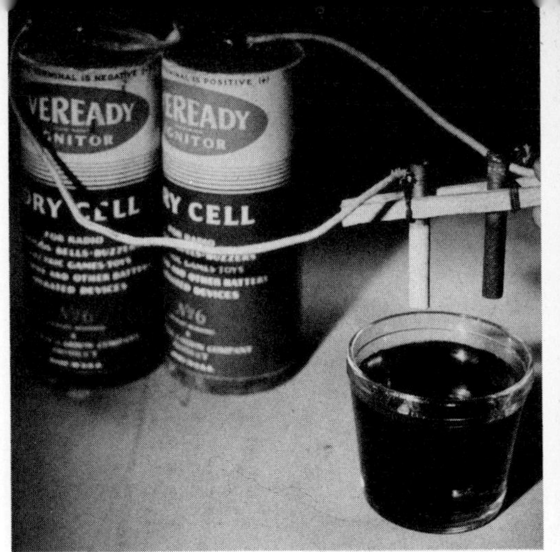

a salt of the metal to be plated and a current is applied. Positive ions of metal are drawn to the cathode and are there deposited as atoms of the metal.

Make a plating bath by dissolving 28 g of copper sulfate in 125 ml of water and adding carefully about 3 ml of concentrated sulfuric acid. For electrodes again use two carbons from old flashlight cells. By clamping these between two small sticks, with the help of rubber bands, you can regulate the flow of current by altering their spacing.

This time two dry cells connected in series should provide about the right amount of current. If the current is too great, the coating of metal is likely to be coarse, and it will not adhere as it should.

Copper soon starts to deposit on the cathode when current is applied. Since the anode is carbon, the copper could come from nowhere except the solution. Within 10 or 15 minutes the coating on the cathode should be heavy enough to take a polish.

**Plating in reverse.** If you want to prove to yourself more definitely the action of the current, reverse the connections to the batteries, the positive lead this time going to the electrode that has the coating of copper. In about the same time that it took to plate the original cathode, the new cathode will become plated. In the meantime, the old cathode will have lost all its copper.

But this does not mean that any of the metal will have been transferred directly to the new cathode. Instead, it will have gone into solution to help replenish the ions changed into metal at the cathode. This is exactly what happens in electroplating when the anode is made of the metal that is being plated. The anode gradually dissolves to help keep the electrolyte up to par.

Finely divided phosphorus, left from solution after the solvent has evaporated, spontaneously ignites and sets fire to paper.

# Fire That Starts Itself

FIRE that breaks forth from a substance without the application of external heat has long been a source of mystery and consternation. Without visible cause, flames burst from an innocent-looking haystack or pile of coal; a wad of cotton waste, soaked with oil, sets fire to a garage; a chemical mixture that ordinarily is harmless suddenly blazes up violently.

Thanks to modern research, chemists now know the reason for almost every type of spontaneous combustion. Barring ignorance, carelessness, or accident, fires starting from this source can be prevented. In most cases the actual visible combustion is just an ordinary fire in which a combustible material has been raised to its ignition point by slow oxidation or other chemical reaction. In the remaining cases, all particularly important to the chemical worker, heat and flame are produced in a reaction involving substances other than oxygen.

**All oxidation produces heat.** Heat is generated every time a substance is oxidized. If the oxidation is extremely fast, we may see fire. If it is slow enough, as in the rusting of iron or the drying of paint, the heat is dissipated so fast that it is impossible to detect any rise in temperature. If, however, we increase the speed of slow oxidation, decrease the rate of heat dissipation, or both, until heat is generated faster than it can be dissipated, we have a starting point for spontaneous combustion.

One way to speed up the oxidation, and at the same time reduce the rate of dissipation, is to break up a substance into smaller particles. The smaller the particles, the larger the surface area to react with oxygen and the smaller its mass to dissipate heat. For instance, iron powder that has been freshly prepared by reduction with hydrogen will heat to the point of glowing when exposed to air, and heaps of fine coal particles are far more liable to smolder and burst into flame than piles of large chunks.

**Spontaneous combustion of phosphorus.** With a tiny piece of phosphorus and a teaspoonful of carbon disulfide, you can present a spectacular demonstration of the effect of subdivision. Unless the room temperature is high, a solid lump of phosphorus left exposed to the air will catch fire only very slowly, if at all. Finely divided, however, it ignites almost instantly.

Although this experiment is perfectly harmless if performed as directed, be extremely careful in handling both of these chemicals. *Keep the carbon disulfide far from any flame, and keep phosphorus, or any solution of it, off your hands, clothes, or anything flammable. Always handle the phosphorus with tweezers or tongs and cut it under water. Dispose of any remaining solution by allowing it to burn in an open pan.*

By gentle shaking dissolve a piece of phosphorus about half the size of a pea in 5 ml of carbon disulfide in a test tube. Use an ordinary cork as a temporary stopper for the tube, as carbon disulfide and its vapors

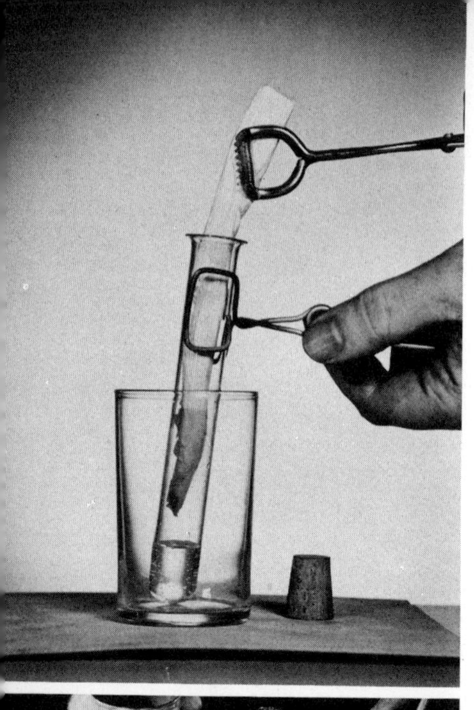

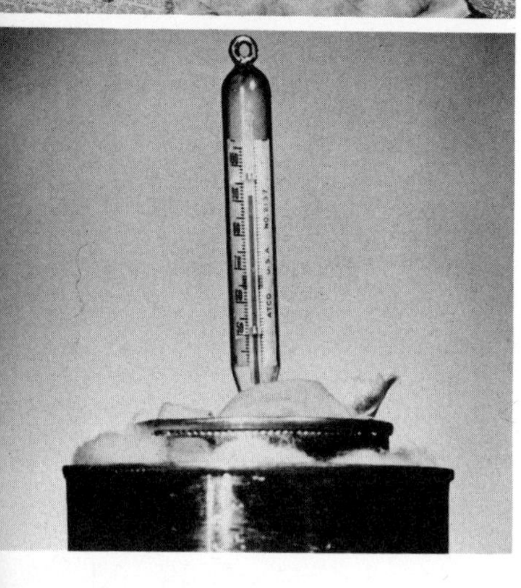

attack rubber. After supporting the test tube in a tumbler so your hands will be free, dip the end of a strip of filter paper or paper towel, held in tongs, into the solution. Remove it, quickly stopper the tube, and hold the paper in the air, away from your face.

Nothing happens as long as any carbon disulfide remains. But this solvent quickly evaporates, leaving finely divided phosphorus on the paper. In a flash, this phosphorus unites with oxygen, heat is generated faster than it can be carried away, and the phosphorus and paper burst into flame.

**Why oily rags sometimes catch fire.** You can also speed up oxidation and reduce heat dissipation by spreading a substance so thinly that its surface is large compared with its mass. Such thin spreading of oxidizable material is the cause of spontaneous combustion in oily rags.

Most vegetable and animal oils, such as linseed, tung, olive, and cod-liver oils, contain linoleic and other fatty acids that react with oxygen at ordinary temperatures to produce the solid substance "linoxyn." As in all oxidations, heat is produced in this reaction. Ordinarily, when such oils are in bulk or are spread in paint over a large open surface, the heat is carried away as fast as it is produced.

At top, paper napkin is dipped in phosphorus solution for first experiment. Middle, cloth is dabbed with linseed oil. Bottom, oxidizing oil gets hot.

**Sulfuric acid steals oxygen and hydrogen from carbohydrates so avidly that often heat enough is produced to start a fire.**

But spread such oils on the fibers of rags or waste and confine the rags or waste so the heat cannot escape, and the story is different. Little by little, the heat increases until finally the ignition point of the cotton is reached and the cotton and oil go up in flames.

You can easily prove that drying oils generate heat. Place a small tin can inside a larger one and insulate one from the other with cotton, asbestos wool, or sawdust. Next paint a few pieces of cotton cloth with dabs of ordinary paint or a drying oil. (Oxidation takes place more rapidly if all the pores and fibers are not completely covered with oil.) Then pack these paint-dabbed rags around the bulb and stem of a thermometer until the small can is full. Finally, set the apparatus in a moderately warm place and note the thermometer reading periodically. The temperature will rise as the oil unites with oxygen from the air.

**Many chemicals can start combustion.** In the chemical laboratory, there are many substances that may react spontaneously to cause combustion. That is one reason why home and student chemists should never mix chemicals together hit-or-miss, to see what happens. When performing an unfamiliar experiment, you should follow instructions to the letter.

It would take several volumes to describe all the chemical reactions that might produce fire or at least considerable heat, but you can easily demonstrate a few.

**Sulfuric acid chars wood.** All home chemists are familiar with the heat produced when sulfuric acid is added to water. (It is because of this that water must never be added to the acid, for the water, being lighter than acid, would remain on top, where it would be heated to boiling by the strong reaction, with the result that hot water and acid might spatter explosively out of the container.) But how many know that heat is also produced when the concentrated acid is dropped on wood, paper, or textiles—enough heat, sometimes, to start a fire? This acid has such a great affinity for water that it removes hydrogen and oxygen, in the exact proportions in which they occur in water, from such carbohydrates, leaving a residue of carbon.

As a demonstration of this, place a drop of acid on a piece of paper, cotton cloth, or wood. In each case the substance will become charred.

**Sugar and acid start a fire.** Some substances that normally are stable when mixed may react violently if a third substance is added. Potassium chlorate and powdered sugar are such a combination. Mix carefully 1 g of powdered potassium chlorate with 2 g of powdered sugar. (Never *grind* potassium chlorate with sugar or any other organic chemical, as the heat of friction may cause an explosion. However, you may grind either chemical *separately* in a clean mortar and then stir them together gently on a piece of paper.) After the two substances are mixed, place about a fourth of the mixture in a little heap on an inverted can cover that, in turn, rests on a table covered with newspapers to catch any possible spatterings.

Then add just one drop of concentrated sulfuric acid to the center of the heap and quickly stand back. In a second, the sulfuric acid unites with the potassium chlorate, forming chloric acid. This oxidizes the sugar so rapidly that the mixture bursts into flame.

**Even water** will react strongly enough with some substances to start a fire. Mixed in the right proportion with ordinary quicklime (calcium oxide), for example, the heat produced has been known to ignite wood. To verify this heating effect, put a teaspoonful into a small crucible, packing it around a strip of paper. After 5 minutes, remove the paper and you will find at least part of it charred.

Ether is formed when two molecules of alcohol lose one of water. Sulfuric acid acts as a catalyst in helping to remove the water.

# Catalysts—Secret Agents of Chemistry

BY means of a catalytic oil-cracking process, industrial chemists now produce a superfuel which is superior to any gasoline previously available. Newpapers, technical journals, and schoolbooks tell us how catalytic processes help make sulfuric and nitric acids, ammonia, dyes, alcohol, plastics, synthetic rubber. In hundreds of other processes we seldom hear of, some catalyst is equally vital. Just what *is* a catalyst, and how does it work?

The anwer to the first part of the question is easy, but the second poses a problem which chemists have never solved satisfactorily. A catalyst is a substance which can change the speed of reaction between two or more

Water is often a catalyst. Here a single drop, added to powdered iodine crystals and powdered aluminum, is sufficient to make the mixture burst into colorful flame.

other substances, without being permanently changed or used up. Jokingly it has been called a "chemical parson"—an agent which can bring about the union of other substances and yet remain unchanged itself.

**Water as a catalyst.** Sometimes the speed of reaction can be accelerated by a catalyst to such a tremendous degree that the catalyst seems to *produce* the reaction. Take a little mound of baking powder, for instance. As long as this powder remains perfectly dry it might remain almost indefinitely without decomposition. If you add a few drops of water, however, the seemingly inert powder immediately froths up with volumes of gas. Plain water has in this case acted as a catalyst, helping the acid compound and the sodium bicarbonate to unite, liberating carbon dioxide. The water is not changed by the reaction.

**Water starts a fire.** Water is probably the most universal catalyst. Many chemical reactions apparently cannot take place without at least a trace of moisture. You can easily demonstrate a violent and vivid reaction between normally nonreactive materials, initiated by a single drop of water. Because of the dense, choking fumes produced, it is best to perform this experiment outdoors, or near an open window.

Powder some iodine crystals in a mortar and mix thoroughly with this powder an equal volume of powdered aluminum, first making certain that both substances are absolutely dry. Place a little mound of the mixture on the center of an asbestos pad. Now add a drop of water in a depression on top of the mound, and stand back. Almost instantly, purple-red vapors rise from the heap. In another second, the mound bursts into beautiful purple-red flame, accompanied by great volumes of smoke.

**Sulfuric acid helps make ether.** Berzelius, the great Swedish chemist, devised the term "catalytic agent" in 1836. Long before that, however, catalysts had been used. Sulfuric acid, for example, served as a catalyst in making ether—or "sulfuric ether," as it is sometimes still called. Ether—or more properly, ethyl ether—is produced from ethyl alcohol when two molecules of the alcohol lose one molecule of water. Sulfuric acid helps remove the water.

To demonstrate this transformation in your home laboratory, pour about 2 ml of denatured alcohol into a test tube and carefully add an equal amount of concentrated sulfuric acid. Pour in the acid very slowly, mixing the solution cautiously with a glass stirring rod. Now heat the mixture gently over a small alcohol or bunsen flame. Keep the tube turned away from you, and do not bring the mixture to a boil. This experiment is entirely harmless if you follow directions; the cautions are merely to prevent any spilling or spattering of the strong acid, which might cause serious burns to yourself or damage to your workbench.

After heating the acid-alcohol mixture for a few seconds, pour it into about 20 ml of warm water in a beaker. The presence of ether can immediately be detected by its smell. The sulfuric acid is merely diluted by the water removed from the alcohol during this reaction. None of the acid is otherwise altered or lost.

**Catalyst produces oxygen foam.** A common chemistry experiment shows how manganese dioxide serves as a catalyst to release oxygen from hydrogen peroxide. Here is a method of demonstrating this process more graphically. Pour about 1 in. of 3 per cent hydrogen peroxide into a tall glass or cylinder and mix with this a few drops of some liquid detergent. Now drop in a few grains of powdered manganese dioxide. Instantly, oxygen released from the decomposing peroxide starts making foam, and if the glass is not more than 10 in. high, the foam will finally overflow it.

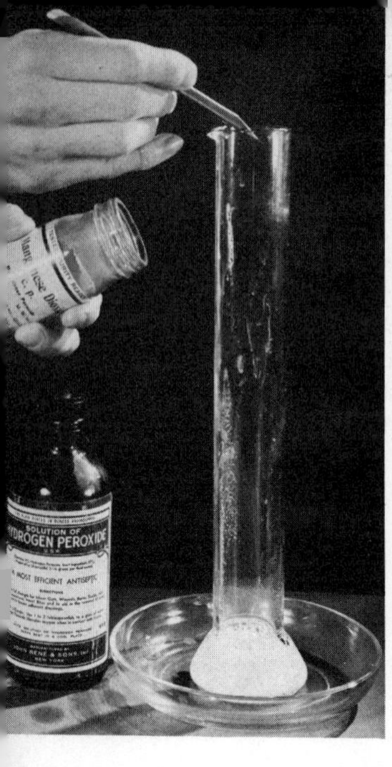

When a catalyst, a few grains of manganese dioxide, is dropped into a cylinder holding some hydrogen peroxide, a rapid decomposition of the peroxide ensues. If a few drops of detergent are added, the liberated oxygen produces foam that climbs over the top.

**Surface catalysts.** The catalysts thus far mentioned have been powders or liquids which have been mixed with the substances between which they have promoted a reaction. In another type of catalyst the reaction occurs when gases or other substances are caused to unite by being strongly adsorbed together on the surface of some metal, such as spongy platinum, platinum black, copper, or finely divided iron or nickel.

**Flameless cigarette lighters** depend for their operation on catalysts of this type. In these lighters, alcohol vapor oxidizes vigorously upon touching the surface of a platinum or palladium catalyst. The intensity of the reaction causes the catalyst to heat up to redness.

**Poisoned catalysts.** When lighters of this kind fail to work properly, the trouble often springs from a "poisoned" catalyst, a condition produced by an accumulation of dirt or soot on the surface of the catalytic metal. (Because an extremely small quantity of an impurity can affect the power of a catalyst, chemists sometimes use this fact to make sensitive analyses. For example, very minute concentrations of cyanide can be detected by their effect in reducing the catalytic activity of platinum.) If your flameless cigarette lighter has a catalytic element coated with soot, you can usually renew it by holding the unit for a few seconds in the tip of a blowpipe flame, as shown on the next page.

A surface catalyst is the secret behind the flameless cigarette lighter. If this catalyst becomes coated with soot, or "poisoned," the lighter ceases to work. You can usually burn off the soot by holding the unit for a few seconds in the tip of a blowpipe flame.

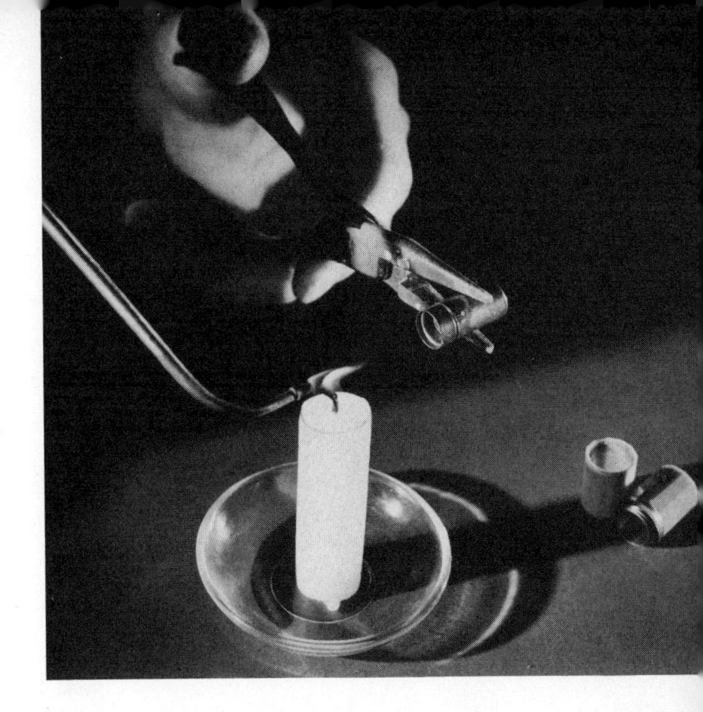

**Copper catalyst makes formaldehyde.** Smooth copper screening is used as a surface catalyst in changing methyl alcohol into formaldehyde by a process of gentle oxidation. This reaction can easily be demonstrated with the aid of a small spiral of copper wire. Procure about eight inches of No. 14 or 18 bare copper wire (bell wire with the insulation removed will do) and form the end into a spiral by winding it around a pencil.

Gently warm a little methyl alcohol in a test tube. Heat the copper spiral in a bunsen flame to dull redness, and then plunge this coil into the vapor above the alcohol in the test tube. Any coating of black copper oxide which may be on the wire immediately vanishes, and the pungent odor of formaldehyde issues from the tube. The hot copper causes the alcohol vapors to unite with oxygen from the air and from its own oxide.

Although it is easy to describe a catalyst and to demonstrate that it works, the question of *why* it works still remains unanswered. Many theories have been advanced to explain catalytic action in specific cases. In some cases it is thought that the catalyst may actually become changed in some intermediate stage of a reaction, only to be regenerated at the end of the reaction. Fame, and possibly fortune, awaits the chemist who can solve the mystery completely.

Mineral oil is broken down into gas and gasoline in this home-lab cracking plant. The resulting vapors burn with a brilliant flame.

# Hydrocarbons—Chemical Building Blocks

FEW of us may recognize the hydrocarbons by their family name, but they are really the most familiar, plentiful, and useful of organic chemical compounds. As the chief constituents of coal, petroleum, and natural gas, they supply nearly all the fuel used for heat and power. Like molecular blocks, they form the starting point for synthesizing dyes, drugs, explosives, plastics, Nylon, synthetic rubber, and other modern marvels of the chemical laboratory.

As the name suggests, hydrocarbons are composed of hydrogen and carbon. Because of the ability of carbon atoms to link up with other carbon atoms to form chains, rings, and more complex structures, hundreds of different hydrocarbon compounds are known. A few of the simpler ones can be experimented with in your own lab.

**Hydrocarbons fall into several distinct series,** the chief of which is called the "paraffin series" because many of its components may be found in paraffin. The simplest member of this series is the gas "methane," which consists of one carbon atom surrounded by four hydrogen atoms. Next comes another gas, "ethane," consisting of two atoms of carbon and six of hydrogen. In this regular order—always increasing by one carbon atom and two hydrogen atoms—the series continues until finally we find giant molecules of 60 carbon and 122 hydrogen atoms.

Physical characteristics change uniformly as molecular size increases. The first four members of the series are gases; the next four are components (in variable degree) of gasoline; then follow kerosene, heating oil, Diesel oil, and finally heavy lubricating oils, grease, and paraffin wax.

When crude oil, or petroleum, which contains dozens of hydrocarbons, is heated to about 300°C and passed as hot oil and vapor into the bottom of a high "fractionating" tower, gas and gasoline are drawn off from the upper part of the tower and naphtha, kerosene, gas oil, and heavy oil from progressively warmer lower sections.

New techniques enable chemists to link light gas molecules in the series to make heavier liquids as well as to "crack" heavier members to make lighter ones. Gasoline is produced by both building up small molecules and breaking down big ones.

**How to crack oil.** Though crude oil is generally cracked in industry by the use of heat, pressure, and catalysts, you can demonstrate cracking by the use of heat alone. If you have no crude oil, medicinal mineral oil will do.

Set up your apparatus as shown in the photo opposite. Pour 5 ml of oil into the test tube at left and fill the upper part with loosely packed steel wool. Through a 1-hole cork stopper (the heat and oil will destroy rubber) run one end of a bent glass tube and extend the other end of the tube through another 1-hole stopper to within 1 in. of the bottom of the

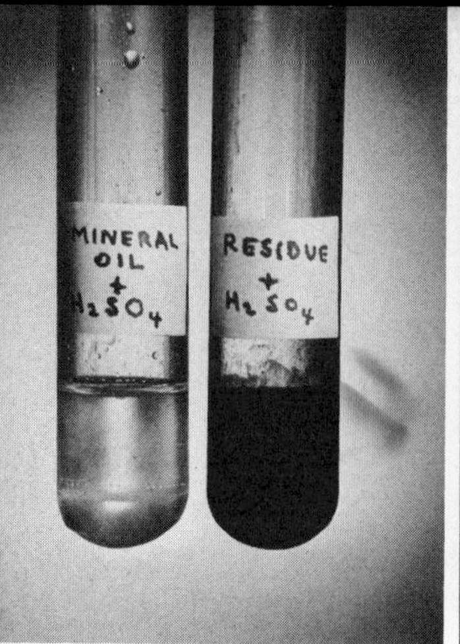


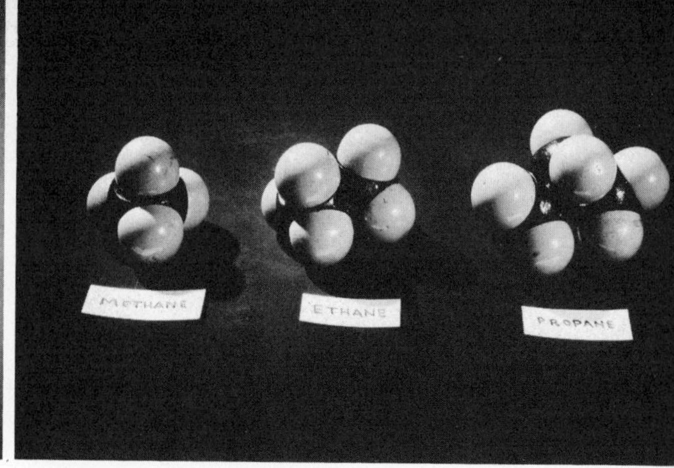


Sulfuric acid has little effect on pure mineral oil, a saturated hydrocarbon, but turns the residue from the cracking process brown. Saturated hydrocarbons belong to the series shown above.

side-necked test tube shown at right. Draw another glass tube into a jet and attach it to the side neck with a short rubber coupling. The wing top flame spreader on the bunsen burner heats both the oil and the steel wool above it.

Start heating gently, and increase the heat gradually. When the oil boils, white fumes will pour into the side-necked tube and some will stream out the jet. In passing through the hot steel wool, some of the molecules of the oil vapor are cracked, or broken down, into smaller molecules much as they are in commercial gasoline cracking plants. A match applied to the jet ignites the mixture of gas, gasoline, and other vapors, and they burn with a brilliant flame.

**Paraffin hydrocarbons are inert.** All the hydrocarbons of the paraffin series are colorless and show little or no reaction with most chemical reagents. In fact, the name "paraffin" comes from the Latin *parum,* "little," and *affinis,* "akin." Shake some pure mineral oil with an equal amount of concentrated sulfuric acid, and there will be no discoloration or other sign of reaction.

One reason for this inertness is that all possible places at which other atoms can combine with the carbon atoms are already occupied by hydrogen atoms. Because of this, members of the paraffin series are known as "saturated" hydrocarbons. When a paraffin hydrocarbon is broken down by cracking, however, there are not enough hydrogen atoms available to fill all the places on the two resulting molecules. One of these, no longer a member of the series, becomes an "olefin," a type of "unsaturated" hydrocarbon.

**One method of forming unsaturated hydrocarbons is by breaking off hydrogen atoms from saturated hydrocarbons. Ethylene can be made by taking two from ethane, acetylene by subtracting four.**

**Olefin hydrocarbons are reactive.** Some of these olefins condense in the bottom of the side-necked test tube when the mineral oil is cracked. When a little of this residue, or condensate, is shaken with concentrated sulfuric acid, the unsaturated olefins react immediately with the acid and become dark brown.

**Methane is sometimes called "marsh gas"** because it is the gas formed by decaying vegetable matter in marshes and stagnant pools. It also issues from coal seams and, mixed with air, has been the cause of mine explosions. Most important is its occurrence as the chief constituent of natural gas.

To show how methane is given off slowly and spontaneously by coal, heap a large handful of finely pulverized coal in the bottom of a large can or jar of water and place an inverted glass funnel over the coal. Attach a glass jet to the end of the funnel by means of a short length of rubber tubing. Prepare the apparatus by sucking up water to fill the funnel tube, and close off the rubber tube with a pinchcock. After several days, the methane gas will rise from the coal and drive some of the water from the funnel. Open the pinchcock and ignite the gas. It will burn with a hot, pale flame.

**Unsaturated hydrocarbons** are also obtained by breaking off hydrogen atoms from members of the paraffin series. The gas ethylene, $C_2H_4$, can be obtained by taking hydrogen from ethane, $C_2H_6$. Ethylene helps to make illuminating gas luminous, helps to ripen fruit, and is an anesthetic and a reducing agent in chemical synthesis. It can be made by heating ethyl alcohol with sulfuric acid.

Ethylene gas can be made by carefully heating a mixture of ethyl alcohol and concentrated sulfuric acid, the reaction being helped by a little powdered pumice or anhydrous aluminum sulfate. The gas is collected for experiments by the displacement of water from jars in your pneumatic trough.

Pour into a 500-ml flask 20 ml of 95 per cent ethyl alcohol (denatured alcohol will do) and 50 ml of concentrated sulfuric acid. Add 10 g of powdered pumice or anhydrous aluminum sulfate to help the reaction. Shake thoroughly, and then support the flask on an asbestos-center wire gauze. Close the flask with a 2-hole stopper having a thermometer in one hole and a bent delivery tube in the other. Heat to 155°C and, after the air has been displaced, collect several tumblers of gas by displacement of water.

**The reducing action of ethylene** may be shown vividly by pouring a weak solution of potassium permanganate, acidified with a drop of sulfuric acid, into a glass of the gas. As the purple solution enters the glass the ethylene robs it of oxygen and so turns it colorless.

The still less saturated gas acetylene, $C_2H_2$, is produced when four atoms of hydrogen are broken off a molecule of ethane or two hydrogen atoms off an ethylene molecule. This is the gas commonly made for camp-lantern use by decomposing calcium carbide with water, and sold in cylinders as the fuel for oxyacetylene torches.

Acetylene enters actively into many chemical reactions. Here acetylene is generated in the presence of chlorine by adding calcium carbide to a mixture of hydrochloric acid and chlorine bleach. As bubbles of acetylene mix with the chlorine at the surface, they burst into flame.

**Acetylene and chlorine produce a flame.** Since the two carbon atoms of acetylene are able to hold four more of some other element, the gas enters actively into many chemical reactions. You may demonstrate this spectacularly by generating it in the presence of chlorine. Make the chlorine by pouring about half an inch of sodium hypochlorite solution (household chlorine bleach) in a large beaker and adding a little hydrochloric acid. Cover the beaker loosely with a piece of cardboard to help retain the heavy gas.

After several minutes, remove the cardboard and sprinkle a little calcium carbide into the solution. Bubbles of acetylene will break through the surface, and as each mixes with the chlorine it reacts so violently that it bursts into flame.

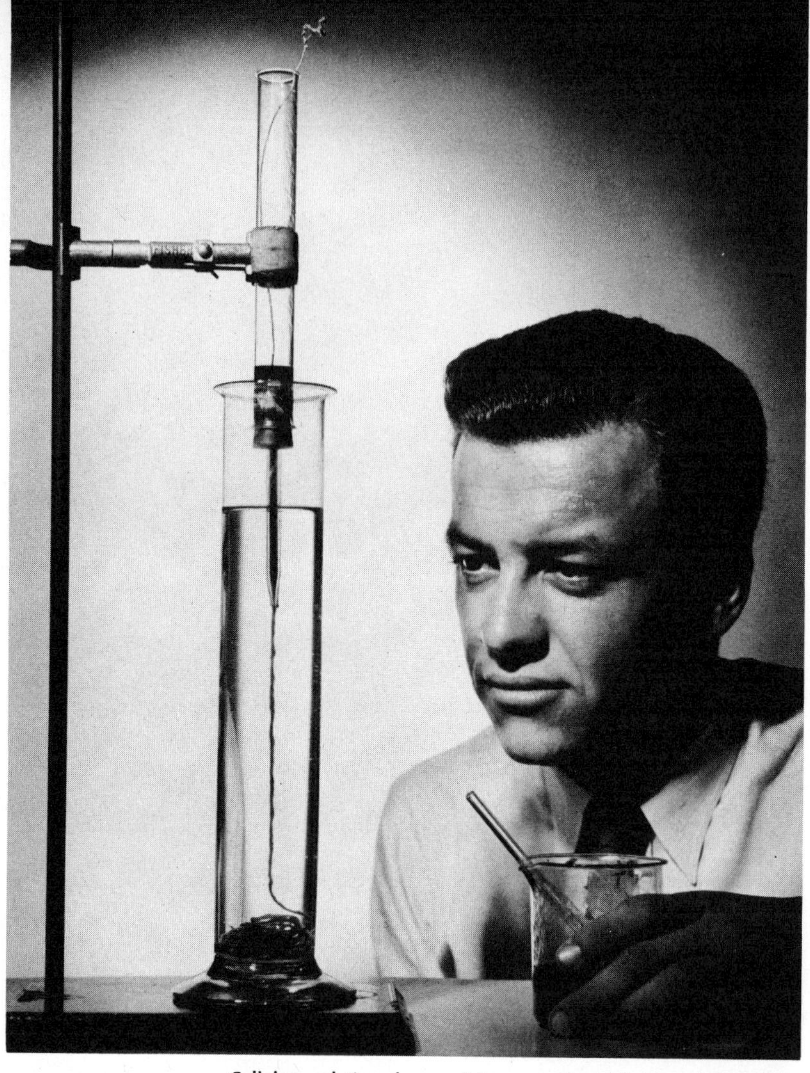

Cellulose solution changes into rayon thread in coagulating bath.

# Chemistry Spins a Yarn

THE transformation of tree fibers or cotton linters into rayon fabrics is one of the greatest achievements of modern industrial chemistry. Chemically, rayon is almost pure cellulose, the same as cotton and linen. Instead of using cellulose as found in nature, however, the rayon chemist starts with cheap plentiful spruce and hemlock trees, or the fuzz which clings to cottonseed after it has been ginned. He chops these up, dissolves them, and then causes the cellulose to reappear in wondrous silky filaments, which may be spun, twisted, knit, or woven, into forms which today compete favorably with cotton, silk, linen, or wool.

**A tip from the silkworm.** Although the rayon industry is comparatively young, the idea that man might successfully imitate the silkworm dates back to the seventeenth century. In his book *Micrographia,* published in 1664, Robert Hooke—minister's son, physician, and scientist—commented on natural silk as being merely a "glutinous excrement" forced out of the body of a little worm. Why, therefore, couldn't men do as well as a worm by making a suitable gum and squeezing it through tiny holes?

To Count Hilaire de Chardonnet goes the honor of first doing this on a commercial basis. Chardonnet was working with Pasteur in trying to find the cause of a mysterious disease that threatened to exterminate the silkworm. He was also interested in photography. One day while he was coating some plates with collodion, the bottle slipped and broke on the table. When he tried to clean up the mess some time later, the partly dried collodion pulled out into long silklike threads. With this as a beginning, Chardonnet worked nearly thirty years to perfect a synthetic yarn. His "Chardonnet silk" was the hit of the Paris Exposition in 1889.

**How to make rayon.** Most rayon is now made by the viscose process. Acetate rayon—now called merely "acetate"—comes next, while cuprammonium, a superior rayon for certain purposes, is in third place. Although complex machinery and precise technical control are required to produce usable rayon yarn, you can readily demonstrate how cellulose in one form can be broken down and then "regenerated" in a form entirely different. The cuprammonium process has been chosen as the one most easily paralleled at home.

First make some copper hydroxide by dissolving 5 g of copper sulfate crystals in 100 ml of water and then adding slowly a 10 per cent solution of sodium hydroxide until the pale-blue precipitate stops forming. Wash this precipitate by decantation—mixing with water in a tall glass, allowing the solid to settle, pouring off the clear upper liquid, mixing with more water, and repeating the process five or six times.

Filter the washed precipitate and dissolve the gelatinous solid in the smallest amount of 28 per cent ammonium hydroxide. With the ammonium hydroxide, the copper hydroxide forms a complex deep-blue compound. The resulting solution (called "Schweitzer's reagent" from the German chemist who discovered it) dissolves paper, cotton, wood, and other forms of cellulose.

To make cuprammonium rayon, add a solution of sodium hydroxide to one of copper sulfate, as at left. Wash and filter the pale-blue precipitate and dissolve it in a little ammonium hydroxide.

**Commercially, cuprammonium rayon** is made from cotton linters. If you wish, you may use absorbent cotton, but filter paper—a pure form of cellulose—will dissolve much faster. Tear up several 15-cm sheets and put these into the blue solution. Stir occasionally until they have completely dissolved. This may take several hours.

In commercial practice, the solution is now filtered and air bubbles are removed from it. Then it is pumped to the spinnerets—disks with tiny holes in them through which the solution is forced. The spinnerets are located at the upper end of a cylinder through which water is flowing downward. This water starts to coagulate the threads and stretches them. The filaments are then completely coagulated in a bath of dilute sulfuric acid.

In your home setup, the "spinneret" can be a glass tube drawn down to a 1-mm opening at one end. The other end of this tube is fitted into a stopper that is inserted, in turn, into the lower end of a tube about 1 in. in diameter and 6 in. long. The large tube serves as a tank from which the solution can be gravity-fed to the coagulating bath, a 5 per cent solution of sulfuric acid in a tall glass cylinder as shown on page 92.

Dissolve enough filter paper in the resulting deep-blue solution to make it syrupy. Allowed to flow through a small opening into dilute sulfuric acid, this solution hardens into a rayon thread.

**Spinning the thread.** Before lowering the outlet tube beneath the surface of the coagulating bath, half fill the tank with the cellulose solution. Be sure the solution is free from undissolved cellulose as this would clog the opening. As a further precaution, run a fine copper wire down from the top of the tank right through the opening. By manipulating the wire, you can clear the opening if it becomes clogged.

To complete the experiment, lower the filled spinneret until the outlet is several inches below the surface of the bath. If the solution doesn't flow, give it a few prods with the wire. At first, the issuing thread may tend to rise, but gradually it will sink. Commercial cuprammonium filaments are much finer, of course, and are stretched during the spinning to give them greater strength. The final treatment consists of washing, drying, and twisting into yarn or thread.

Viscose rayon usually is made from cellulose derived from wood pulp. After being steeped in sodium hydroxide solution, this pulp is squeezed almost dry and allowed to age for several days. Then the alkali cellulose is treated with carbon disulfide, which changes it to cellulose xanthate. Dissolved in sodium hydroxide solution, this forms a viscous amber-

To make acetate rayon, dissolve filter paper in a mixture of acetic acid, acetic anhydride, and sulfuric acid. Pour the solution into a beaker of water to precipitate cellulose acetate.

colored solution that is forced through spinnerets in much the same way as cuprammonium solution.

**How to make acetate.** Unlike cuprammonium or viscose rayon, "acetate" is not cellulose that has been taken apart and put together again. Instead, it is a real compound of cellulose—cellulose acetate, the material of which most photographic film is made. All you have to do to make filaments of this substance is to dissolve it in acetone, or other suitable solvent, squirt it through little holes, and finally pass it through warm air. The heated air rapidly evaporates the solvent, leaving a solid thread.

You can make cellulose acetate by dissolving 1 g of filter paper in a mixture of 40 ml of glacial acetic acid and 12 ml of acetic anhydride, with 6 drops of concentrated sulfuric acid added as a catalyst. Complete solution will take at least 8 hours. Stir it occasionally.

When the paper has dissolved completely, you can recover the cellulose acetate by pouring the solution in a thin stream into a large volume of water, stirring constantly. Then separate the compound by filtering it,

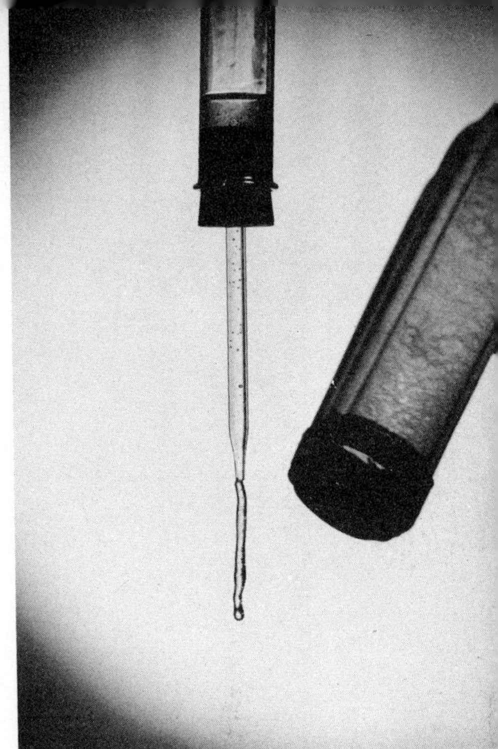

Filter this compound, dry it, and then dissolve it in a little acetone. Acetate thread is "spun" by letting it ooze through a small aperture. Warm air quickly evaporates the solvent.

press out the excess water by squeezing the compound between filter papers, and dry—either in the open air or by directing upon it the warm blast from an electric drier.

When dry, dissolve in the smallest possible amount of acetone. Place the resulting solution in the "tank" used in the previous experiment. This time the solution just flows into the air where heat from the drier solidifies it.

White cotton remains white as long as it is left in a solution of indigo. But remove the cloth, and it mysteriously turns blue.

# Chemistry Robs the Rainbow

FEW chapters in the history of chemistry are more romantic than the story of dyes. Long before he could write, prehistoric man had learned to stain his body and dye his garments with juices extracted from plants and insects. Wrappings on Egyptian mummies buried before 3000 B.C. show clearly the blue dye of the indigo plant and the red of the madder root. A secretion that the ancients obtained, drop by drop, from a tiny mollusk found on the shores of the eastern Mediterranean yielded the famous Tyrian purple, a dye so expensive that it long was the badge of royalty.

All of the early dyes were laboriously obtained from plants, certain insects and mollusks, and a few minerals. Those few that did not fade from washing or sunlight could be counted on the fingers of one hand.

**A boy chemist discovers coal-tar dyes.** A great step forward came in 1856. Working in an attic laboratory in his London home, William Henry Perkin, an eighteen-year-old chemistry student, in that year stumbled upon the treasure of coal-tar dyes. His discovery revolutionized the dye industry.

At the time, Perkin was trying to synthesize quinine from aniline oil, a secondary substance which is produced from the messy black tar left as a waste product when coal is burned for gas and coke. Instead of white crystals of quinine, he got an ugly mass of a sticky dark substance. When he tried to rinse it from the test tube, the mass surprisingly dissolved into a beautiful purple solution.

This result intrigued Perkin and he did not rest until he had found the reason. His purple, or violet, solution was soon named "mauve." This first commercial coal-tar dye was not only the inspiration for the "mauve decade," but was the foundation for a gigantic industry. Within a few years Perkin and other chemists had synthesized magenta. Then came synthetic alizarin—replacing almost overnight a $20-million-a-year madder-root industry—and indigo, the greatest single dye of them all. Today there are more than 5,000 coal-tar dyes, hundreds of which are more varied, brilliant, and more lasting than anything produced by the ancients.

**What makes a dye?** Any colored solution is not necessarily a dye. To serve as a dye, the solution must penetrate a textile or other material and deposit its coloring matter so firmly in the fibers that handling and washing will not remove it. To be really "fast," the color also must not fade or change after prolonged exposure to light and air.

The simplest to use are "direct" dyes, those which attach themselves directly to a textile either by chemical reaction with the fiber or by mechanical adhesion. Many vegetable dyes and most packaged synthetic dyes from the drugstore are of this kind. Especially when used on wool (which contains organic acid and nitrogen groups that react with certain dyes) direct dyes are often brilliant and serviceable.

Many of the most permanent synthetic dyes, however, are developed directly in the textile fiber by means of a few amazing tricks of chemical sleight of hand. Indigo, aniline black, azo colors, and mineral dyes all make use of this magic.

**The magic of indigo.** With a few grams of indigo and some sodium hyposulfite, you can demonstrate the application of what is undoubtedly the best inexpensive blue dye ever made. Sodium hyposulfite is known as "sodium hydrosulfite" in the dyeing trade and should not be confused with sodium thiosulfate, the "hypo" used in photography.

As purchased, indigo is a dark-blue powder or paste that is insoluble in water. To make it soluble, it must be reduced (that is, deoxidized) to a colorless compound. For your demonstration, mix 2 g of powdered indigo with 1½ g of sodium hydroxide dissolved in 5 ml of water in a test tube. Then stir in an additional 20 ml of water and support the tube in a beaker of water heated to 50°C. When the solution is heated, stir it well and slowly add 1½ g of sodium hyposulfite.

Let this mixture stand for about half an hour, keeping the temperature constant and stirring occasionally. At the end of that time, the blue indigo should have changed to a yellow or colorless "indigo white," which, in turn, should have completely dissolved in the solution of sodium hydroxide.

**Indigo is called a "vat dye"** because in days gone by the color was reduced and made soluble by a fermentation process in huge tanks or vats. Today hyposulfite takes the place of these ferments. You can prepare an experimental "vat" by adding a few grains of hyposulfite to 250 ml of water in a beaker and heating the water to 50°C. After 10 minutes, add the contents of the test tube. The solution should now be pale yellow and clear.

Now for your magic. Dip a white cotton cloth into your vat and let it remain for a minute. As long as it is submerged, the cloth stays white. But lift it out and in a minute it turns blue. Repeated dipping and exposure darken it.

What happens is this: By reaction with the oxygen of the air, the deoxidized colorless indigo saturated through the cloth is changed back into indigo of the original color. Because indigo is insoluble, washing will carry away only the little that is attached loosely to the surface. The remainder is bound permanently to the fabric.

**The wonder of aniline black.** Another important and extremely fast coal-tar dye is aniline black, a dye developed directly in the cloth by the oxidation of aniline. The process makes an easy but startling chemical demonstration. To perform it, you will need a little aniline, the colorless oily liquid which in the hands of chemists yields dyes of every hue of the rainbow. (*Be careful in handling this chemical, for the fumes and the liquid are poisonous. If you spill any on your hands, rinse them with water immediately.*)

A cloth wet with aniline hydrochloride solution is normally colorless. But hang it in steam, and it is dyed aniline black.

Prepare some aniline hydrochloride by adding 3 ml of concentrated hydrochloric acid to 2 ml of aniline in a small beaker. Dissolve the resulting crystals by adding 10 ml of water. To this solution add 3 g of potassium chlorate, as an oxidizing agent, and 1 g of copper sulfate to serve as a catalyst.

Soak a strip of white cotton cloth in this solution for several minutes. When you remove it, it will still be white. But blot off the excess solution and suspend the strip in a flask in which a little water is boiling strongly, and the strip will change mysteriously to black. The heat and moisture of the steam hasten the oxidation of the aniline, changing it from a colorless compound to a beautiful washproof dye.

**Magic with mineral dyes.** A number of excellent mineral colors, considerably older than coal-tar dyes, also are made by precipitating an insoluble chemical directly into a textile. Chrome yellow, a brilliant color, is a good example.

In one glass dissolve ½ g of lead acetate in 25 ml of water, and in another ¼ g of potassium bichromate in an equal amount of water. Dip a strip of cotton in the first solution, remove it, blot off the excess solution, and dip in the second. Yellow lead chromate is precipitated into

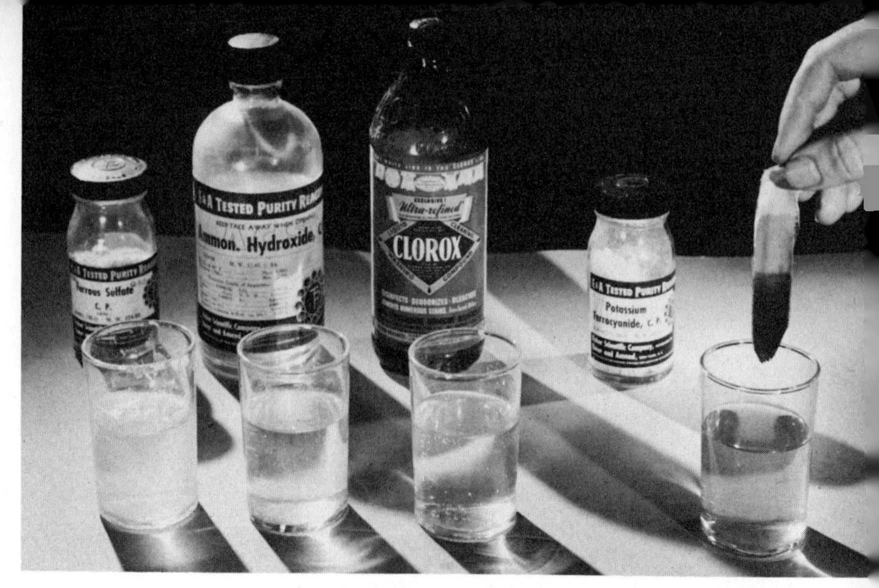


These four solutions, all of which are colorless, will produce either iron buff or prussian blue, two important mineral dyes.

the cloth. Repeat the process and the color becomes deeper. You can change the color to the bright orange of basic lead chromate by passing the dyed cloth quickly through a boiling solution of calcium hydroxide.

Another widely used mineral color is iron buff, the basis for some khaki dyes. This is really iron rust deliberately precipitated into a textile. You can produce the color by dipping a cloth successively into three colorless solutions. First pour 100 ml of water into each of three glasses. In the first, dissolve 2 g of ferrous sulfate; in the second, 5 ml of concentrated ammonium hydroxide; in the third, 10 ml of 5 per cent sodium hypochlorite solution.

Now dip your strip of cloth, blotting it between dips. The ammonia will change the ferrous sulfate into green-gray ferrous hydroxide. The hypochlorite in turn oxidizes this into brown ferric hydroxide, which, when dry, changes into the oxide. A deeper color can be made by repeating the sequence.

By more chemical magic you can change your iron buff into prussian blue. Dip the dyed cloth into a solution of 1 g of potassium or sodium ferrocyanide and 20 drops of hydrochloric acid dissolved in 100 ml of water. This solution itself will be colorless, but when the buff-dyed cloth is taken out it will have changed to an attractive blue.

**Mordants hold dyes fast.** Many dyes that ordinarily fail to attach themselves strongly to certain textiles can be made to do so by first treating the textiles with colorless insoluble compounds called "mordants."

Cloth painted with two colorless substances known as "mordants" (left) becomes dyed in two colors when dipped in a single dye.

These usually are metal hydroxides which cling to the textile fibers and either absorb or react chemically with the dye. When a chemical reaction takes place, the resulting color is called a "lake" and may be entirely different from that of the dye. Designs of different colors sometimes are produced with a single dye and different mordants.

You can show this by painting part of a piece of cotton cloth with a solution of ferrous sulfate, such as that used in the iron buff experiment, and another part with a similar solution of potassium alum. Leave a third part of the cloth unpainted.

Blot off the excess solution, dip for a minute in a 5 per cent solution of ammonium hydroxide, blot once more, and immerse for 10 minutes in a 2 per cent suspension of alizarin red dye in water. When the cloth has been removed and rinsed, you will find that the part mordanted with the alum will remain a bright orange-red, that the area mordanted with iron will be a dark red-brown, while the part which was not mordanted at all will remain practically colorless.

The first step in making bakelite is to boil the ingredients gently. The condenser returns high-boiling vapors to the flask.

# Polymerization—Big Word in Synthetics

DON'T let this tongue-twister scare you. "Polymerization"—the key to a whole wonder world of synthetic plastics, rubber, gasoline, and other creations of the modern chemical laboratory—is really easier to under-

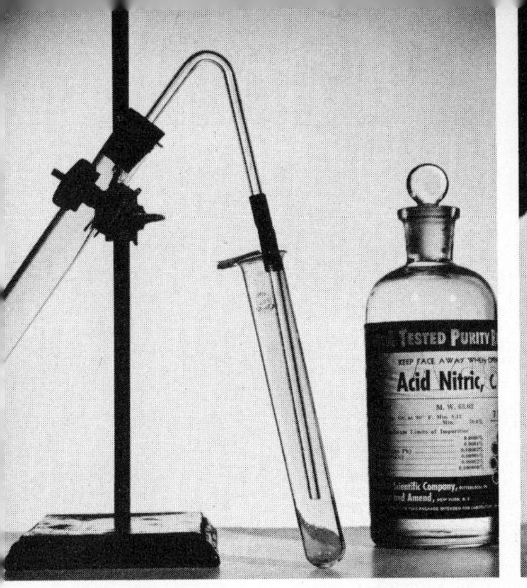

**A tube of nitrogen dioxide gas in hot water becomes darker; one in ice water becomes colorless. Generator for gas is at left.**

stand than it is to pronounce. "Polymer," its root, comes from two Greek words meaning "consisting of a number of parts." "Polymerization," then, means the chemical grouping together of a number of molecules of one substance, or of related substances, to form bigger molecules.

Natural polymers—starches, proteins, cotton, silk, wool, leather, and wood—are as old as creation. By studying these, chemists have learned both how to modify such natural materials and to synthesize new ones. Rayon, cellophane, photographic film, and cellulose lacquers, for instance, are made by taking apart the cellulose of wood and cotton and putting it together in new forms. Bakelite, Lucite, Nylon, and buna rubber, on the other hand, are examples of chemical synthesis, products different from anything in nature.

Heat, cold, light, mechanical mixing, and catalysts may alone or together take part in this molecule building. The result often is little short of miraculous. Gases lose or gain color, or change into liquids. Liquids become thicker liquids, flexible resins, or even solids. Some substances which ordinarily are highly reactive to heat and chemicals are transformed into heat-stable materials that resist powerful chemicals.

**Cold changes light gas to heavy gas.** As an introduction to this process, let's try a graphic experiment in which a colored gas becomes a heavier colorless one merely by the application of cold.

Mount a test tube on a stand as shown above, drop in several short pieces of copper wire, and add several milliliters of concentrated nitric acid. Close the tube at once with a stopper having a delivery tube leading to the bottom of another test tube standing vertically. Place a bit of

Here are the materials for making bakelite. After boiling them together, the resulting liquid is poured into a tin can cover.

cardboard loosely over the mouth of the vertical tube to help retain the gas that will form.

This reddish-brown gas is nitrogen peroxide, a mixture of nitrogen dioxide and nitrogen tetroxide (*Caution: Be careful not to inhale this gas*). When the color indicates that the receiving tube is full, close it tightly with a rubber stopper and collect a second tube of gas in the same manner.

If you now stand one tube in a glass of hot water and the other in a glass of ice-cold water, you will witness elementary examples of both polymerization and depolymerization. The gas in the heated tube becomes darker, more of its molecules breaking down into nitrogen dioxide ($NO_2$), while that in the chilled tube becomes almost colorless, its molecules doubling up to become nitrogen tetroxide ($N_2O_4$).

Now put the cooled tube into the hot water and the warm one into the cold water. The reaction is immediately reversed. Both compounds, $NO_2$ and $N_2O_4$, have the same elements in the same proportions, but the polymer $N_2O_4$ is twice as heavy as the other.

**Liquid changes to solid.** Formaldehyde is a gas, but it usually is sold as a solution in water. If you let some of this solution stand in the open in a slightly warm place, a strange thing happens. As the water evaporates, a white solid is left behind. By polymerization, an indefinite number of formaldehyde molecules joined with water and formed the

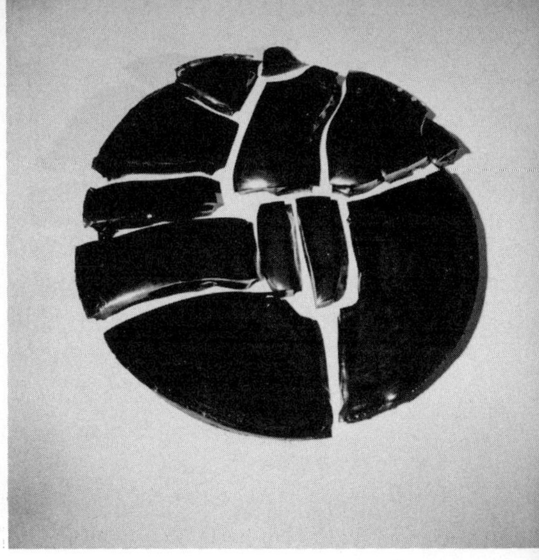

Polymerization is completed by baking the liquid as shown at left. Upon cooling, the disk may crack, but it is real bakelite.

solid—paraformaldehyde. If you heat this solid, it changes back to gas.

**You can make bakelite.** An important product of polymerization is bakelite, a synthetic resin that is made by reacting phenol with formaldehyde. During the reaction, a water-thin, water-clear, highly reactive solution is transformed into a ruby-red, glassy solid that will withstand heat and the effects of most chemicals. If you have a condenser, and running water to cool it, you can demonstrate this wonder in your own lab.

Make your starting solution by dissolving 38 g of phenol (carbolic acid) in 75 ml of 40 per cent formaldehyde solution. (*Be careful not to spill the phenol on your hands, for it is caustic and poisonous. If you should accidentally spill it, wash it off immediately with plenty of water.*) Next add about 5 ml of a 40 per cent sodium hydroxide solution to act as a catalyst.

Carefully pour this combined solution into a 300-ml boiling flask, put a wire-gauze square under the flask, and mount it on a ring stand. If you own one, connect a reflux condenser to the mouth of the flask through a 1-hole stopper as shown on page 104. If you lack one of these, an ordinary condenser can be substituted. This arrangement allows steam to escape but returns substances with a higher boiling point to the solution. Connect the jacket of the condenser so that water flows in at the bottom and out at the top.

**Solution thickens and colors as it boils.** Now start the solution boiling gently. The total boiling time necessary will probably exceed an hour, but take a look at the flask every few minutes to observe the strange change that takes place.

If your phenol is absolutely pure, the solution at the beginning will look like water. If it is impure, the solution will have a pinkish tinge. In either case, as the boiling continues, the solution turns a light amber, gradually becoming darker and darker with a trend toward red. The solution also gradually thickens.

Because of differences in conditions of boiling, the time required to complete the reaction cannot be accurately predicted, but must be judged by observation. Keep a constant watch after boiling has continued for about 50 minutes. As soon as the solution thickens to the approximate consistency of molasses, turn off the flame, remove the condenser, and quickly pour the solution into a mold. (*Be sure to turn off the flame as the first step, for the vapors from the flask are flammable.*) Do not allow the solution to boil beyond the point indicated or it will solidify in the flask. As soon as the flask has been emptied, clean it with a brush and a strong solution of trisodium phosphate or of lye.

**Baking completes the transformation.** The cover from a large tin can may be used as a mold. If you allow the solution to cool at this point, however, you will discover that you have a resinous substance that melts again on slight heating and dissolves in many organic compounds. To complete the process of polymerization, you must bake the resin several hours at a moderate temperature.

This can be done by putting the mold, with the solution in it, on an asbestos mat and suspending over it a 100-watt bulb in a photographic reflector. A thermometer supported directly over the mold will indicate the approximate temperature. For the first hour, adjust the temperature to about 50°C, then raise it to 75° for the following 2½ hours. When the polymerization is completed, the substance will be a deep red.

During your cooling, your bakelite will shrink and may crack because of internal strains, but it will come out of the mold a shiny, brittle, glass-like disk of ruby red.

To prevent subsequent shrinking, commercial bakelite is molded and baked under high pressure. A filler such as clay, asbestos powder, or wood dust often is mixed with it to reduce its cost, and pigment may be added to give it a color other than red.

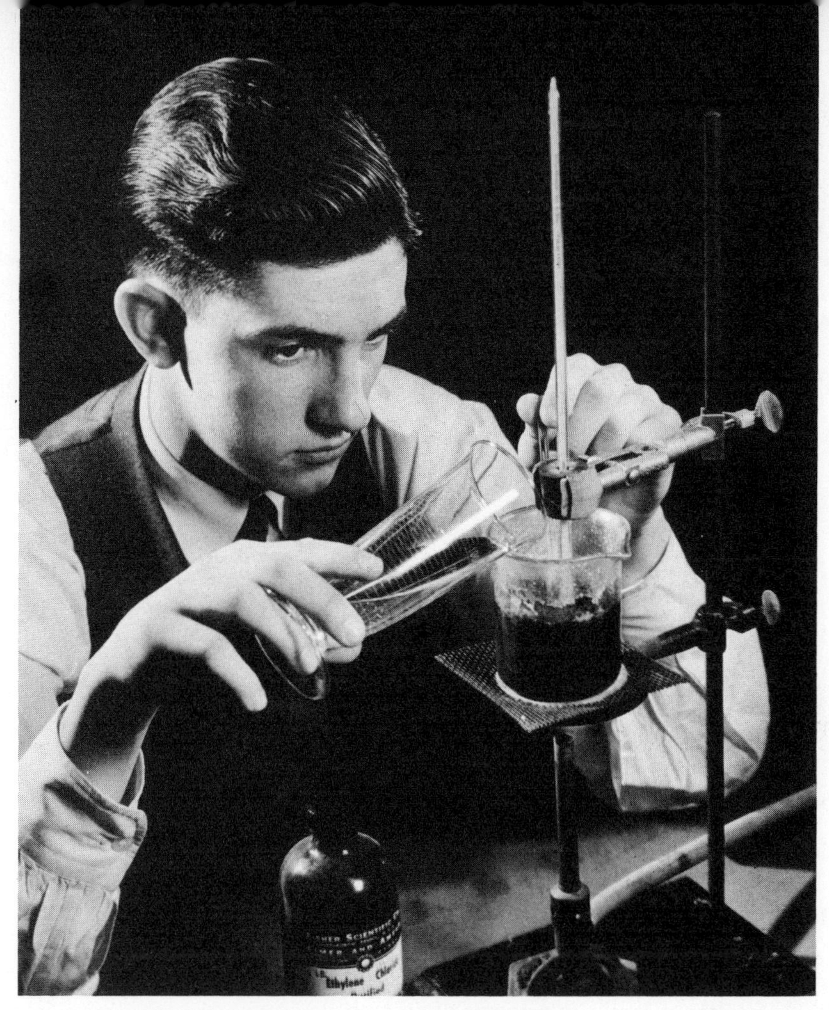

One of the last steps in synthesizing rubber is to add ethylene dichloride to sodium polysulfide while watching the temperature.

# Synthetic Rubber Chemistry

IF the ancient sage who contended that you "can't make a silk purse from a sow's ear" could glimpse a few of the modern miracles worked by chemical synthesis, he might change his mind and decide that almost anything is possible. Materials even more commonplace than pigs' ears are now constantly being transformed into completely different substances of far greater value than their constituents.

For instance, polysulfide rubber, a widely used rubber substitute, utilizes just a few simple chemicals. You can make it right in your own

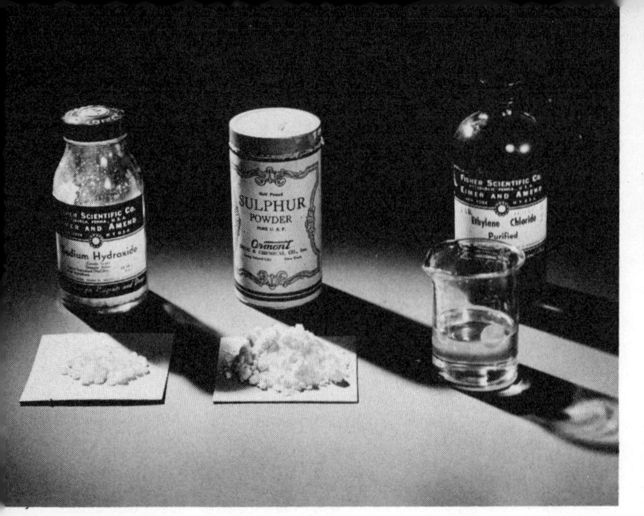

The starting materials for making Thiokol are shown at left. As the first step, dissolve sodium hydroxide in water in a beaker.

laboratory at home. It is doubtful whether your experiments will yield new tires for your car, but the product that you obtain will look and feel like natural rubber, and, if you wish, it can be turned into a ball or used as a pencil eraser.

**Under the trade name of Thiokol,** polysulfide rubber was the first type of synthetic rubber produced in the United States. Not only has this synthetic all the bounce and stretch of natural rubber, but it is far more resistant to oil, gasoline, and other organic solvents.

What's it made of? Just a dab of sodium hydroxide (lye), a dash of sulfur, and a dribble of ethylene dichloride (a liquid no more exciting-looking than water)—all chemicals that have about the same relationship to the finished raw rubber as the sow's ear does to milady's silk pocketbook.

**Little molecules into big molecules.** The objective in making all types of synthetic rubber is to cause the small molecules of thin solutions to link themselves together and form big molecules of rubberlike consistency. This process, described more fully in the previous chapter, is known as "polymerization." Sodium polysulfide and ethylene dichloride provide the small molecules that become linked together when polysulfide rubber is produced.

**Your first step in synthesizing rubber** is to make a quantity of sodium polysulfide. Measure 150 ml of water into a beaker and dissolve in it 10 g of sodium hydroxide (ordinary lye will do). Heat this until it

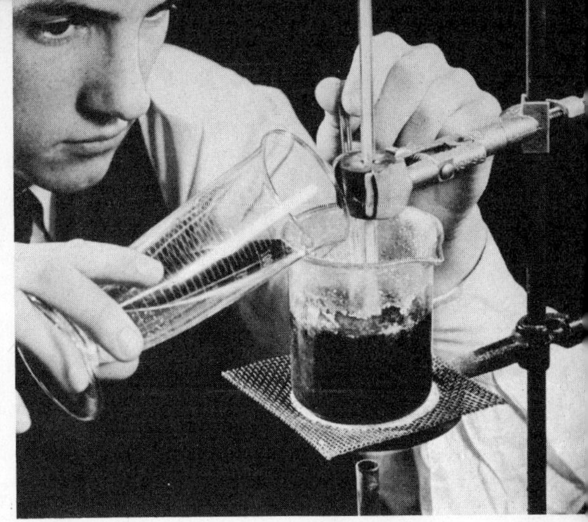

**Bring the solution to a boil, and make sodium polysulfide by adding sulfur. To the clear liquid, add ethylene dichloride.**

boils. Then add slowly, while stirring vigorously, about 20 g of flowers of sulfur. This substance is hard to wet, but persistent and vigorous stirring will win out. The solution will change gradually in color from light yellow to dark brown as the sulfur reacts with the sodium hydroxide. This darkening is an indication of the increase in the sulfur content of the polysulfide. The reaction will have reached its limit after 10 minutes of boiling and stirring. Then you should allow the solution to settle and cool. Finally, pour off the dark-brown liquid into another beaker, leaving the unreacted sulfur behind.

**Ethylene dichloride,** sometimes merely called "ethylene chloride," is the second major constituent of polysulfide rubber. It is an oily organic liquid that does not mix in the ordinary sense with any water solution. Therefore, it can be held in contact with the solution of sodium polysulfide only in the form of a physical suspension produced by a constant and vigorous agitation.

Such physical suspension can be attained a little more easily by using a few grains of some chemical known as a "dispersing agent," a compound that helps to keep groups of particles of substances apart by neutralizing their cohesive attraction. In this experiment magnesium hydroxide can be used as the dispersing agent. Add it to the solution of sodium polysulfide before proceeding to the ethylene dichloride step.

**Make your own magnesium hydroxide.** If you do not have magnesium hydroxide on hand, you can make this compound by adding a solution of sodium hydroxide to a solution of magnesium sulfate (epsom salts)

As the mixture cools, particles settle out. Washed and treated with acid, they coagulate into a solid mass of synthetic rubber.

until precipitation stops. The precipitate, which is only slightly soluble in water, settles slowly. After it has fully settled, pour off the upper clear liquid and add fresh water. To wash the chemical, repeat this settling and pouring-off process several times. A white powder will remain. This is the magnesium hydroxide you need.

Stir a little of this powder into your solution of sodium polysulfide and reheat the solution to about 70°C, checking the temperature with a thermometer supported in the beaker. Keep the thermometer in place throughout the remainder of the experiment and watch the temperature closely so that it does not get too high.

**Now comes the ethylene dichloride.** Slowly add about 30 ml, stirring constantly and very actively to disperse the liquid throughout the sodium polysulfide solution. The ethylene dichloride starts reacting with the sodium polysulfide immediately. Since this produces additional heat the thermometer must be watched carefully to keep the temperature of the mixture from rising above 80°C, which is close to the boiling point of the organic liquid.

Continue stirring the mixture. If the particles are not well dispersed, lumps of a yellowish soft material may be encountered by your stirring rod. The color slowly changes from a transparent dark brown to an opaque lighter shade of brown. Keep up the stirring for 15 minutes. Then turn off the heat and let the mixture stand.

As it cools, minute particles of a yellowish rubbery material will settle slowly to the bottom. When the mixture has fully settled, pour off the upper clear liquid, as in making the magnesium hydroxide, and refill the

Squeezed into a firm ball, the synthetic will bounce almost as well as natural rubber, and you can stretch it considerably before it breaks. You can put it to practical use as a pencil eraser.

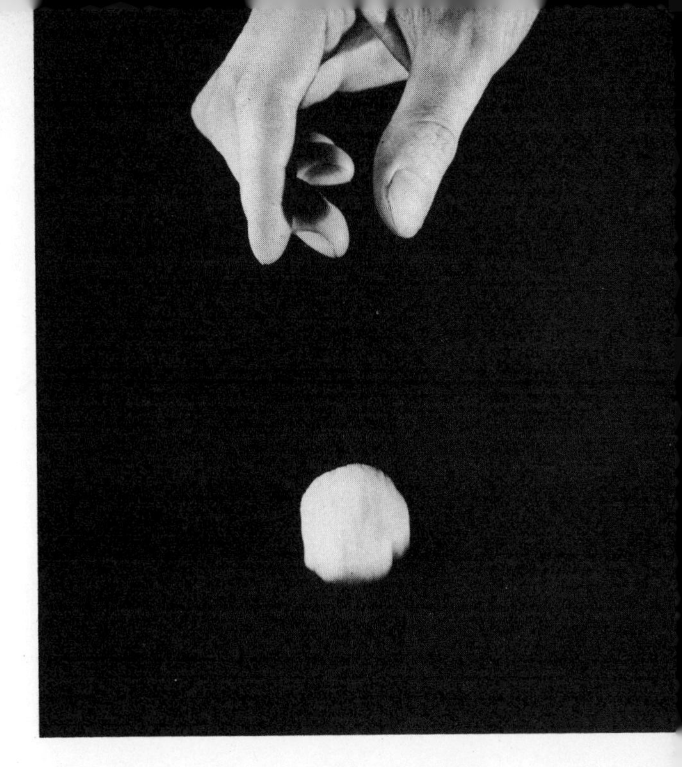

beaker with plain water. Repeat, as before, until the water is almost colorless.

**Acid helps coagulation.** The material at the bottom by this time may have coagulated into a mass. If it has not, add several drops of phenolphthalein solution. Because of the presence of a little alkali remaining, this will cause the liquid to turn pink. Now add dilute hydrochloric acid, drop by drop, until the pink color disappears. After the color has gone, add several more drops to make the liquid slightly acid. This acidification of the solution breaks down the repulsion between the particles of polysulfide rubber and causes them to coagulate into one irregular mass.

**Your rubber is finished.** Take out the mass, rinse it, and squeeze it into a ball. What you have is a blob of real, raw synthetic rubber with about the consistency of an eraser. It will bounce, although not too well, and you can stretch it considerably before it breaks.

Commercially, this raw synthetic is used just as the starting point for producing finished rubber for many different purposes. It can be worked on a rubber mill in the same manner that natural rubber is worked. If the raw plastic mass is intimately mixed with zinc oxide and heated to about 140°C, a transformation occurs similar to the vulcanization of natural rubber.

Air in ordinary tap water makes it taste better but spoils many solutions by oxidizing them. This apparatus shows its presence.

# How Chemistry Softens Water

BY means of tests and chemical reactions easily duplicated at home, water chemists—the men who stand guard over the drinking and industrial water supply of the nation—make a unique contribution to the saving of human lives, materials, and machinery.

Because water is such a good solvent, it is never found absolutely pure in nature. Even while falling as rain, it takes dust, oxygen, nitrogen, and carbon dioxide from the air. As it runs over rocks and through soils, water picks up mineral compounds that give it various degrees of hardness, and organic matter that colors it, gives it an odor, and too often serves as a breeding place for germs.

But with its simple tricks, chemistry changes all this. Hard water is changed to soft; smells, color, sediment are removed; germs are killed.

Chemicals added to river water (left) carry down dirt (center). Filtering the top portion results in crystal-clear water (right).

Many municipalities and industrial establishments have elaborate water-conditioning plants which are kept under the constant supervision of chemists.

**Chemicals remove suspended matter.** One of the commonest jobs is the removal of suspended matter. Filtering through beds of sand and gravel helps accomplish this. But as too much sediment would clog the filters, most of the suspended particles are removed first by precipitation with alum, or aluminum sulfate.

You may demonstrate this with a glass of water having a little ordinary soil stirred in it. A few drops of a solution of alum and one of lime added to the murky water will cause a precipitate of white, gelatinous aluminum hydroxide to form. Stir the mixture, then let it stand, and in a short time the aluminum hydroxide will catch most of the suspended particles and carry them to the bottom.

**Filters remove color and odor.** If the water is naturally soft, this is followed by filtering through sand, gravel, and sometimes charcoal, the last of which removes color and odor. To show this, color some of the clear water from the last experiment with ink, and pass it through a paper filter partly filled with activated charcoal. The charcoal should remove most, if not all, of the ink by adsorbing it.

**Soft and hard water.** Besides suspended dirt and coloring matter, water may contain the dissolved salts of calcium, magnesium, and other elements. If it reaches you with less than 100 ppm (parts per million) of such mineral salts, water is said to be "soft." If there are more, it is called "hard."

**Hard water is costly.** Since mineral salts form scum in wash water and crusts inside boilers and pipes, hard water has long caused a lot of inconvenience and costly damage. Soap usually consists of a salt of sodium with some organic acid, such as stearic or oleic. Dissolved in water containing calcium or magnesium, soap reacts to form insoluble, sticky, and curdy stearates or oleates of these metals. Until enough soap is added to react with all the hardness, the soap has no power to clean.

In heating systems, the effects of hard water may be still more serious. When water containing carbonate hardness is heated, an insoluble scale is deposited inside the pipes and boilers. Even in a thin layer, this scale reduces the transfer of heat to such an extent that the boiler metal may become overheated and corrode excessively or crack. Layers sometimes become so thick in pipes that the flow of water is seriously impeded—or even stopped.

**Water containing carbonate hardness** is sometimes called "temporary" hard water, because mere boiling is sufficient to soften it. Such water owes its hardness to dissolved calcium bicarbonate and occasionally smaller amounts of the bicarbonates of magnesium and iron. These are produced by the reaction of water containing carbon dioxide with the normal carbonates of these metals. When such hard water is boiled, the bicarbonates are decomposed, forming insoluble calcium carbonate, iron carbonate, and magnesium hydroxide. These settle out, leaving the water soft.

**How to make temporary hard water.** The reactions that cause the formation and destruction of carbonate hardness are easily demonstrated. Shake a few specks of precipitated chalk (calcium carbonate) in about 10 ml of distilled water in a test tube. Like pure water flowing over limestone rock in nature, this water dissolves virtually none of the chalk, but forms with it a milky suspension that gradually settles out.

In a second test tube, shake an equal amount of chalk in water that contains carbon dioxide—carbonated soda water, or distilled water through which carbon dioxide has been bubbled for several minutes. This time the chalk dissolves and a clear solution results, duplicating what happens in nature when rain water charged with carbon dioxide flows over limestone. The water in this tube now has carbonate hardness.

Various carbonates dissolve much more readily in water containing carbon dioxide, as shown in the left-hand photo. Such water is said to have temporary hardness because boiling, as in the other photo, drives out the carbon dioxide and so precipitates the carbonates.

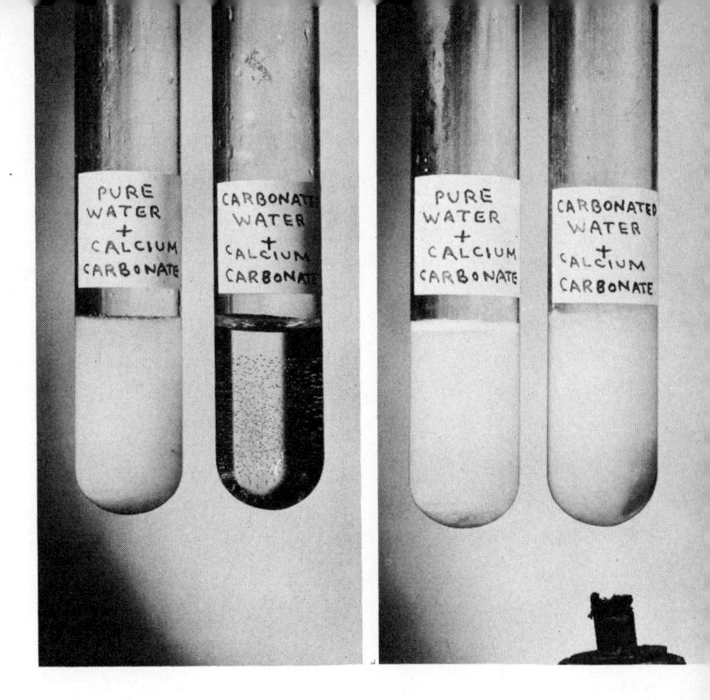


As a demonstration of how such hardness may be removed, heat the water in this tube to the boiling point. The excess carbon dioxide will be quickly driven off and the clear water will be whitened by a precipitate of the original normal calcium carbonate, which will settle to the bottom.

**"Permanent" hard water** generally contains sulfates of calcium or magnesium, which are more difficult to remove. The most common method is to treat it with washing soda, soda ash, or crude sodium carbonate. Plaster of paris is a form of calcium sulfate, and epsom salt is magnesium sulfate. Add a little of either to water in a test tube, and then add a little washing soda. The soda will change the salt to an insoluble carbonate. What remains in solution is sodium sulfate, which will not harm boilers or interfere with washing.

**In complete water-conditioning plants,** river or lake water is first mixed mechanically with alum, lime, and sodium carbonate, which remove suspended matter and precipitate any salts that make water hard. The water is next filtered and led to a reservoir for treatment with chlorine gas to kill bacteria. Then as clean, soft, safe water, it goes to the consumer.

**How to test your tap water.** Chemists constantly test water under their care for the nature of its ingredients in order to know how to remove the harmful ones. In your own lab, begin by analyzing tap water in a test tube. Add a little ammonium chloride, ammonium hydroxide, and ammonium oxalate. Boil for a minute. A white precipitate indicates calcium in the form of calcium oxalate.

Now test for magnesium by filtering off the calcium oxalate, if any was precipitated, and adding to the filtrate a little more ammonium chloride and ammonium hydroxide. Add next a few grains of sodium hydrogen phosphate, and shake the tube. The formation of a white crystalline precipitate—ammonium magnesium phosphate—indicates magnesium.

Iron may be present in the form of ferrous or ferric compounds. Add a few drops of hydrochloric acid and a little potassium *ferro*cyanide to a sample. A dark-blue precipitate (prussian blue) shows *ferric* salts. Add hydrochloric acid and potassium *ferri*cyanide, and a somewhat similar dark-blue precipitate signifies iron in the *ferrous* form.

Lead, dissolved for instance from pipes, may be detected by adding a few drops of hydrochloric acid and of hydrogen sulfide solution to a sample of suspected water. A black or brown coloration means lead.

**Air gives water most of its taste,** and air causes most of the rusting action on metals and oxidizing of such substances as photographic developers. The large amount of air that is dissolved invisibly in water may be proved by gently heating tap water in a flask to which an inverted test tube is connected, as shown in the photo on page 114. Fill both flask and test tube completely with water; the bent tube will permit some to flow out as heat expands it and as air is expelled. Bubbles of air rise through the flask and test tube and collect at the top of the latter. Water so heated has the flat taste of boiled water. The taste may be restored merely by whipping air in again with an egg beater.

**The presence of nitrates** often indicates that water may be contaminated by animal refuse. Here is a test which is delicate: to about ten parts of nitric acid to a million parts water. To about half an inch of a water sample in a test tube add an equal amount of concentrated sulfuric acid and *cool.* Then, carefully and without mixing, add a strong solution of ferrous sulfate. The formation of a brown ring where the two liquids join indicates a nitrate.

Does the baking powder in your kitchen comply with government standards for carbon dioxide? This apparatus will let you know.

# Tests for Household Products

WITH prepared foods, medicines, cosmetics, cleaning preparations, and other household products purer and more consistent in quality now than a generation ago, thanks are due to a vast army of chemists working behind the scenes. Chemists working for the manufacturers check products constantly for purity and uniformity. Chemists of the Federal Food and Drug Administration double-check to protect the public from harm and misinformation.

Using common items in your pantry, laundry, and medicine cabinet, you can duplicate many of these tests. Dozens of ordinary household products, such as baking powder, ammonia, beauty preparations, ice cream, soft drinks, tea, coffee, and hydrogen peroxide, are chemical compounds whose mysteries you can probe in your home lab.

**Carbon dioxide in baking powder.** Baking powder—always a mixture of at least two chemical compounds—makes an interesting start. Its effectiveness in raising cake and biscuits lies in its ability to liberate carbon dioxide gas. According to government standards, baking powder must give off at least 12 per cent of its weight as available carbon dioxide. You can check your baking powder with a simple test.

Set up your apparatus as shown in the photo on the preceding page. A tube to the bottom of a small Erlenmeyer flask is connected with a funnel above it by means of a rubber tube provided with a pinchcock. A bent delivery tube leads from the flask to the mouth of an inverted cylinder, which is supported just under the surface of the water in a pneumatic trough. Fill the cylinder completely with water at the start.

Put exactly 2 g of the baking powder to be tested into the flask, insert the stopper, and close the pinchcock. Now measure out 50 ml of water and put it in the funnel. Make sure there are no air bubbles in the stem of the funnel; then open the pinchcock, allow all the water to run into the flask, close the pinchcock quickly, and shake the flask gently for about a minute. Carbon dioxide gas will bubble up into the cylinder. If the cylinder is graduated, the volume of gas can be observed directly; if it is not, just mark the level of the gas in the cylinder, fill the cylinder to this point with water and then measure the water. After deducting 50 ml from this volume to allow for the air forced out of the flask by the water, multiply the remainder by 0.002, which is the approximate weight, in grams, of 1 ml of carbon dioxide. From this you can readily calculate the percentage of gas liberated by the baking powder.

One ingredient of all baking powders is bicarbonate of soda. The others are compounds that form acids in the presence of water. The commonest of the acid-forming substances are tartrates (tartaric acid or potassium acid tartrate), phosphates (calcium phosphate), and alum (sodium aluminum sulfate). A few baking powders contain all three, others contain one or two. To find out which your brand has, stir 10 g of baking powder in 50 ml of water until foaming has stopped. Then filter and test the filtrate.

For alum, put a little of the filtrate in a test tube and add a few drops of a solution of barium chloride. A white precipitate, which will not dissolve when dilute hydrochloric acid is added, is a sure sign of the presence of a sulfate such as sodium aluminum sulfate.

Acidify another portion of the filtrate with a few drops of nitric acid. Warm this mixture and add to it several drops of 10 per cent ammonium molybdate solution. A bright-yellow precipitate denotes a phosphate.

**You can find the relative strength of two samples of ammonia, turned pink by phenolphthalein, by adding acid drop by drop.**

Phosphates in other common substances such as trisodium phosphate cleansers and soft drinks can be tested by the same method. The bright precipitate is sometimes used as a pigment.

To test for tartrates, pour a few milliliters of the filtrate into an evaporating dish, add several drops of concentrated sulfuric acid, and evaporate completely over a small flame. If tartrates are present, the residue will be charred and give off the odor of burnt sugar.

**How strong is your ammonia?** You can find the relative strength of two samples of household ammonia—which is really a solution of ammonia gas in water—by means of titration (see page 158). Pour 5 ml of each solution to be tested into separate test tubes or small beakers, and add a drop of phenolphthalein solution to each to turn it pink. Now make a solution of 1 part sulfuric acid and 4 parts water. (*Caution: Always add the acid to the water, not the reverse!*) Add this drop by drop to each sample, agitating the tube as you do so, until the solution just turns water-white again. The number of drops required to neutralize each sample indicates the relative strength of the ammonia. Use only clear ammonia in this test, as the cloudy variety contains soap and other misleading ingredients.

**Household "peroxide"** is also a solution—hydrogen peroxide in water—and if not tightly stoppered, the hydrogen peroxide may break up into

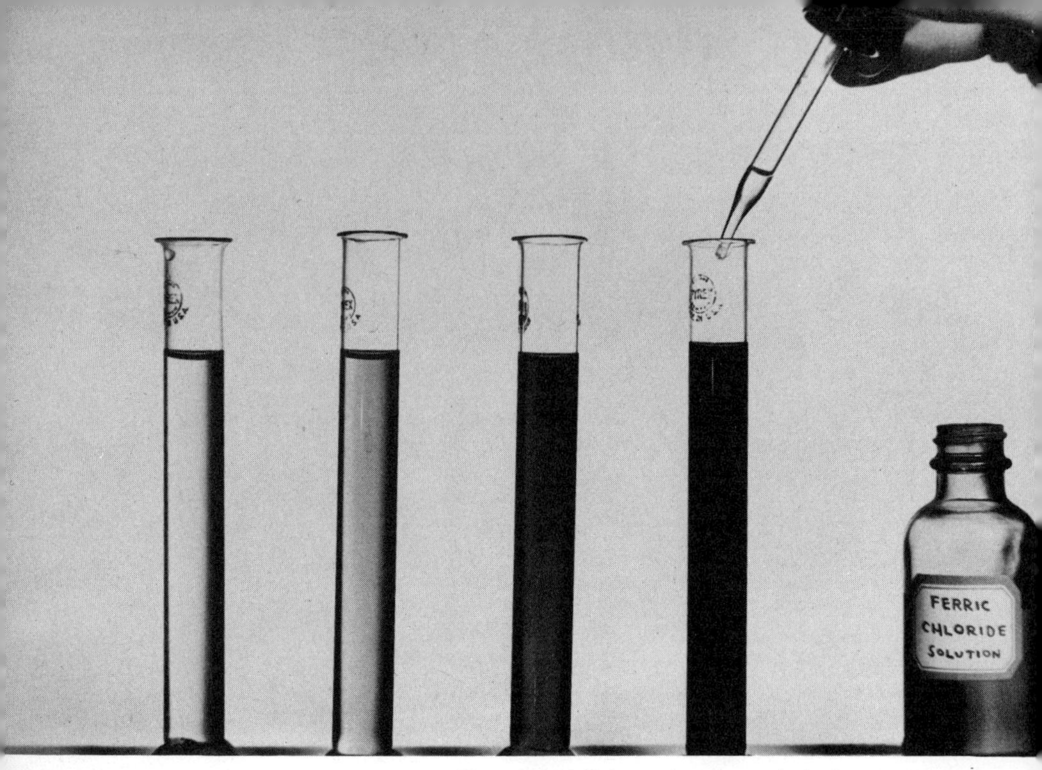


**That tannic acid in tea increases with the length of time the tea is brewed is shown in this graphic test using iron chloride.**

oxygen and water and be worthless. To test a questionable sample, add several drops of dilute hydrochloric acid, and then add a little dilute solution of potassium bichromate. If hydrogen peroxide is present, a blue color will develop. This usually fades quickly, but a little ether shaken in will dissolve the blue solution and form a stable layer on top.

**Gelatin** is rarely, if ever, added to milk nowadays to give it body, but a frequent legitimate use is in ice cream, to increase its smoothness. To test for it, make a solution of acid mercuric nitrate by dissolving a little mercury in twice its weight of concentrated nitric acid (*careful!*) and diluting the resulting solution with 10 times its volume of distilled water. Add to 10 ml of melted ice cream 20 ml of the mercuric nitrate solution and 40 ml of water. Let it stand for a few minutes and then filter. If gelatin is present the filtrate will be slightly cloudy.

**Tannic acid in tea.** That continued brewing increases the tannic acid in tea can be demonstrated with samples brewed for 1, 2, 3, and 4 minutes. To each add iron chloride solution drop by drop until no more black iron tannate precipitate forms. Now shake the tubes. Each successive brew will be darker, indicating more tannic acid.

**If commercial glucose is added to honey,** an iodine test will disclose its presence. Dilute a small sample with an equal amount of water and add a few drops of a solution of 1 g of iodine crystals and 3 g of potassium iodide in 50 ml of water. When commercial glucose is contained in the honey, the color will turn red or violet because of erythrodextrins ("red" dextrins) that are almost always found in the artificially manufactured product.

**The presence of boron** in such products as tooth powder, borax, and boric acid may be demonstrated vividly by means of the setup shown on the right. A glass tube of small diameter passes through the stopper of the test tube and extends about a quarter of an inch into the wider tube clamped above it. This combination acts like a bunsen burner, permitting air to mix with vapors that issue from the test tube below.

The sample to be tested is put in the test tube and covered with a few milliliters of denatured alcohol. Add several drops of concentrated sulfuric acid to act as a catalyst. Then stopper the test tube, adjust the tube above it, and bring the solution to a boil. After a few seconds, light the vapor which pours from the upper tube. If boron is present, the vapor will burn with a vivid green flame.

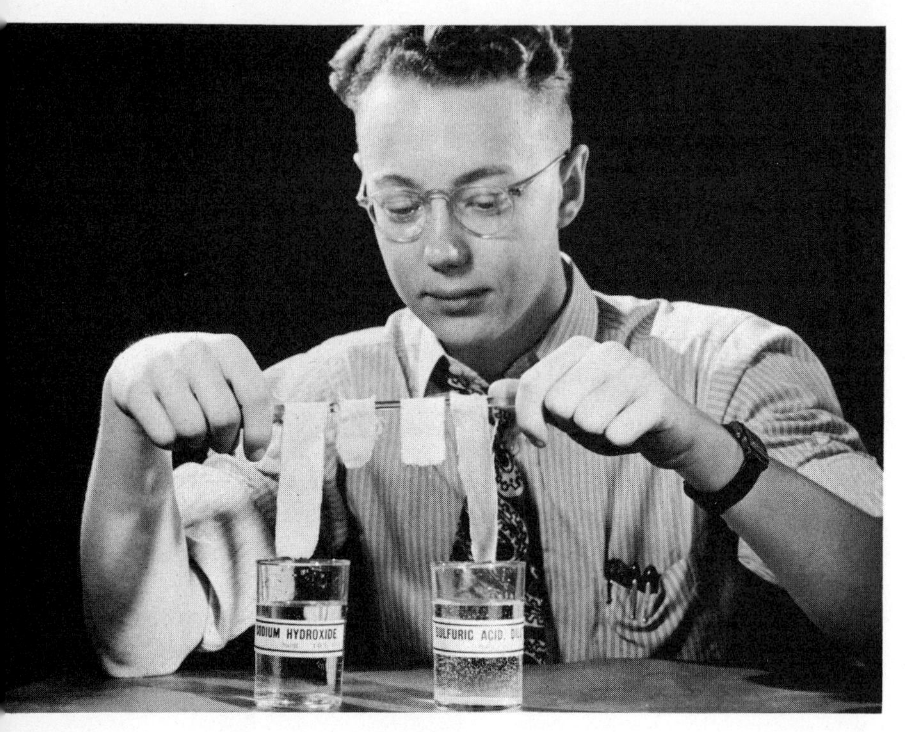


Dip cotton and wool in sodium hydroxide, and the wool disappears; put like strips in sulfuric acid, and the cotton is eaten away.

# The Chemistry of Spot Removal

ONLY a few years ago, when all cloth consisted of natural fibers colored with natural dyes, cleaning was chiefly a matter of sufficient soap and water, backed up by plenty of elbow grease, while obstinate spots were usually attacked by rule-of-thumb measures that had been handed down for generations.

Today, however, an ever-widening array of rayons and other synthetic fibers such as Nylon and Dacron, plus man-made dyes of every imaginable hue and chemical composition, have raised textile cleaning to a scientific industry. Organic solvents have taken the place of soap and water on many fabrics. Workers in spot-removal rooms of the bigger cleaning plants have become textile experts and chemists.

Several organizations of cleaners and dyers maintain laboratories where scientists are searching constantly for better methods of cleaning fabrics. The processes and procedures they have evolved are recorded in a series of excellent manuals.

Although you would need one of these books, plus considerable experience, to handle many of the problems that confront the professional cleaner, you can begin exploring this subject by using only the chemicals and skills you already have on hand.

Before treating textiles with chemicals, however, you should first learn what the common ones can and cannot tolerate.

**Cotton and wool react differently.** A simple experiment will show the general reaction of vegetable and animal fibers to mineral acids and strong alkalies. Vegetable fibers include cotton, linen, and ordinary regenerated cellulose rayon. Animal fibers include wool, silk, and hair.

Suspend strips of cotton and woolen cloth beside each other in a glass containing a 10 per cent solution of sodium hydroxide, and hang a similar pair of strips in a warm solution of dilute sulfuric acid. Examine the strips after ½ hour. You will find that the sodium hydroxide has disintegrated the wool, while the acid has eaten away, or at least badly weakened, the cotton. The other strips of wool and cotton are scarcely affected by the solutions.

Here, then, is striking proof that woolens never should be placed in water containing strongly alkaline washing powders (made up of such constituents as sodium carbonate and trisodium phosphate) and why mineral acids should be kept well away from cotton fabrics.

**Acetone dissolves acetate.** Acetone is an excellent solvent for paints and lacquers on almost all textile materials, but cleaning chemists will warn you never to use it on acetate rayon, or "acetate." Put acetone on acetate —and that's the end of the material. It dissolves in this liquid as readily as sugar does in water. You can easily identify acetate by burning the end of a small sample. It melts as it burns, leaving a hard black knob of ash.

**Colored textiles** admittedly are the most difficult to tamper with, for a solvent or a bleach that will remove a stain on such material may often remove or alter the color of the original dye as well. Before attempting to remove a spot from colored goods, you should therefore first test your treatment on an inconspicuous part of the material.

**Chemicals cause color changes.** One form of coloration change in a dyed material comes from an alteration in the chemical nature of the dye.

**Never treat synthetic textiles with cleaning solvents without first testing the solvent on a small sample. Put acetone on acetate rayon, for example, and the photo at left shows what will happen. The rayon dissolves!**

Basic dyes, for instance, often change color when touched with a weak acid such as might derive from fruit juice or perspiration. A red basic dyestuff, known as "congo red," gives a startling example. Touched with acid, this red dye turns a vivid blue. Instead of discarding a garment that has so changed its color, you can change the color back again merely by rinsing out the acid and neutralizing the spot with a weak alkali such as ammonia.

Acid dyes that have changed color by contact with alkalies often can be restored in a similar manner by rinsing and treating with 28 per cent acetic acid. Provided that they do not affect the dye, acetic acid and ammonia (as a weak acid and a weak base respectively) may be used on virtually all textile materials.

**Spots or stains on white material** generally are removed by employing one or all of the following three methods:

1. Solution in a solvent.
2. Adsorption of a fresh stain by a finely powdered substance.
3. A reaction which bleaches the stain.

In selecting a solvent, a knowledge of chemistry is highly important. To pick the right one you must know not only the nature of the spot but what chemical agent will dissolve it.

Plain water will wash out such things as blood, sugar, and animal glue. Greases and other substances such as candle wax, chewing gum, adhesive tape, and shoe polish require an organic solvent. Carbon tetrachloride and benzol are two such solvents. Because benzol is highly flammable, whereas carbon tetrachloride doesn't burn at all, the latter has become one of the most common solvents used in household spot removal.

**A general paint remover.** Paint varies in its composition, but cleaning chemists have developed a formula for what they call a "general paint remover." It consists of equal parts of benzol, carbon tetrachloride, and amyl acetate (banana oil). The first two ingredients remove ordinary paints, while the last takes care of lacquers. If the amyl acetate is pure, this formula may be used for almost all textiles. Impure amyl acetate may damage acetate rayon.

**Adsorption often may be used effectively** on fresh stains, reducing the color and making final cleaning easier. This procedure consists of moistening the spot with water or organic solvent, depending on the nature of the stain, and spreading a powder over it. Part of the material causing the stain is then adsorbed by the powder. Precipitated chalk is excellent for this purpose.

**Reactions which bleach stains** either give oxygen to the compound causing the stain or take oxygen from it. In either case the compound is changed from a colored to a colorless one.

Ordinary hydrogen peroxide and sodium hypochlorite solution (Chlorox, Rosex, and the like) are the most common of the oxidizing bleaches now used. Stains from fruit juices, coffee, tea, and slight scorching generally respond to peroxide of the 3 per cent drugstore variety.

Sodium hypochlorite solution is much faster than hydrogen peroxide, but it should be used only on cotton, linen, or synthetic cellulose fibers. A little acetic acid added to the solution will make sodium hypochlorite work even faster by liberating chlorine. When sodium hypochlorite is used, the area later should be treated with an "anti-chlor," or chlorine remover. This can be a solution of sodium bisulfite, acetic acid, or ordinary photographic hypo.

As previously stated, sodium hypochlorite is a faster oxidizing agent than hydrogen peroxide. You can determine their relative speeds by adding a little to separate containers of dilute ink or dye and timing the reaction.

**Potassium permanganate,** however, is by far the strongest and fastest of the oxidizing agents—so strong and fast, in fact, that it is rarely used

**A dye-spotted cloth is about to be dipped in potassium permanganate solution (left). Dipping makes the cloth look hopeless (center). But dip it in oxalic acid, and it comes out white!**

except on the most obstinate spots. Strangely enough, the 1 or 2 per cent solution employed for bleaching is a deep purple color, often far more conspicuous than the spot it helps to remove.

Pour a little of the solution into a shallow container, such as a saucer, and immerse in it a bit of white cloth spotted with ink or dye. Remove the cloth after a few seconds—and it is all stained purple. Rinse it in water and it remains an ugly brown. If you now dip the cloth in a 10 per cent solution of oxalic acid (*poison!*) or of sodium bisulfite, however, both the permanganate and the original stain will disappear. The cloth is left whiter than before.

Permanganate solution can also be used for bleaching out tobacco stains from the fingers. Swab a 2 per cent solution on the stained spots. Let it remain a few seconds and then wash your fingers with the sodium bisulfite solution.

**Sodium bisulfite or oxalic acid** sometimes is used alone to remove the color from stains, for each acts as a "reducing agent"—a substance which *robs* oxygen from the colored compound. If cleaners find they can't remove a stain by oxidizing it, they try reducing it. Between the two, the stain generally disappears.

## HOW TO REMOVE COMMON STAINS

Because modern textiles and dyes react so differently to chemical treatment, always test an inconspicuous piece of a material before attempting to remove a spot. After applying water-soluble chemicals, always rinse a textile thoroughly to remove traces of reaction products.

**Acids.** In colored fabrics, moisten with household ammonia.

**Alkalies.** Soften the material with water. Moisten colored fabrics with vinegar; white fabrics with a ½ per cent solution of hydrochloric acid.

**Blood.** Wash fresh stains with lukewarm water. Soften older ones with household ammonia and then treat with a 2 per cent oxalic acid solution.

**Coffee and cocoa.** Wash in concentrated salt water.

**Copper sulfate.** Moisten with 10 per cent acetic acid, then with a 20 per cent solution of sodium chloride.

**Fruit.** Apply sodium bisulfite solution, slightly acidified with hydrochloric acid. Wash well in cold and warm water.

**Grass.** Treat fresh stains with warm alcohol; old ones, with sodium perborate solution. For obstinate stains use a chlorinated bleach followed by a sodium thiosulfate solution.

**Iodine.** Wash with sodium thiosulfate solution.

**Mercurochrome.** Use a 2 per cent solution of potassium permanganate, followed by a 5 per cent solution of oxalic acid.

**Oils and fats.** Place blotting paper under stain. Swab with carbon tetrachloride.

**Rust.** Use a 10 per cent citric acid solution, or a warm 5 per cent solution of oxalic acid to which 5 per cent glycerin has been added.

**Tar and tar products.** Soften the spot with warm oil, then treat like oils and fats.

You can demonstrate the principle of refrigeration with the help of sulfur dioxide gas, made and liquefied in the apparatus above.

# Cold from Chemistry

THE mystery of the household refrigerator that makes cold from electrical energy or the heat of a flame, and of the strange powders that need merely to be dissolved in water to conjure up cold for drinks and compresses, may easily be solved with the help of a few experiments in your home laboratory.

All home-refrigerator systems depend upon a simple law of physical chemistry: every time a liquid evaporates into a gas, it snatches a definite amount of heat from its container and the surrounding air, cooling both below their original temperatures.

The oldest method of cooling by evaporation used plain water as the refrigerant. Primitive man found that water placed in unglazed earthen vessels would seep through the pores, evaporate, and cool the water remaining inside. Campers and country dwellers still cool water in this way.

Your kitchen refrigerator is just a glorified evaporative cooler, having insulated walls to keep out room heat, and pumps, coils, radiators, and control devices to maintain a constant cold inside. In order to freeze ice cubes and keep frozen foods from thawing, a liquid is used in the cooling coils which actually boils at a temperature below the freezing point of water. Common refrigerating liquids are ammonia, sulfur dioxide, and Freon-12 (dichlorodifluoromethane).

**Pressurized spray produces cold.** Because it boils at room temperature, Freon-12 is also used to supply pressure in the household "bug bomb." With such a bomb of the harmless deodorant or insecticide type, you can demonstrate dramatically how spontaneous boiling of this chemical produces cold. Moisten a short piece of glass or metal tubing on the outside with water and hold one end over the nozzle of the bomb by means of a spring clothespin (to keep from freezing your fingers). Then press the release button. Evaporating and expanding as it escapes through the tube, the chemical quickly causes ice to form on the outside.

**How to liquefy sulfur dioxide.** After the Freon, sulfur dioxide, or other chemical in a refrigerator has changed from liquid into vapor, it must be changed back into a liquid again. The electric refrigerator does this by first squeezing the vapor by means of a compressor to concentrate its heat, and then discharging the heat into the room through radiator fins. For purposes of demonstration, you can make a liquid refrigerant by directly cooling sulfur dioxide gas below its boiling point.

Set up your apparatus as shown in the photo on the facing page. The flask is provided with a 2-hole stopper which is fitted with a short bent delivery tube and a dropping funnel whose stem nearly touches the bottom. (You can improvise a dropping funnel, as shown, by connecting the top of a broken thistle tube to a straight glass tube by means of a short rubber tube, using a screw clamp on the rubber tube to control the rate of dropping.) Put 15 g of sodium bisulfite into the flask, and a little dilute hydrochloric acid (1 part acid, 2 parts water) into the closed funnel. To dry the gas, fill the large horizontal tube with lumps of calcium chloride, packing cotton loosely in each end before inserting stoppers with entry and exit tubes. Lead the exit tubes into a large test tube supported in the freezing mixture, and plug the mouth of the test tube with another loose wad of cotton.

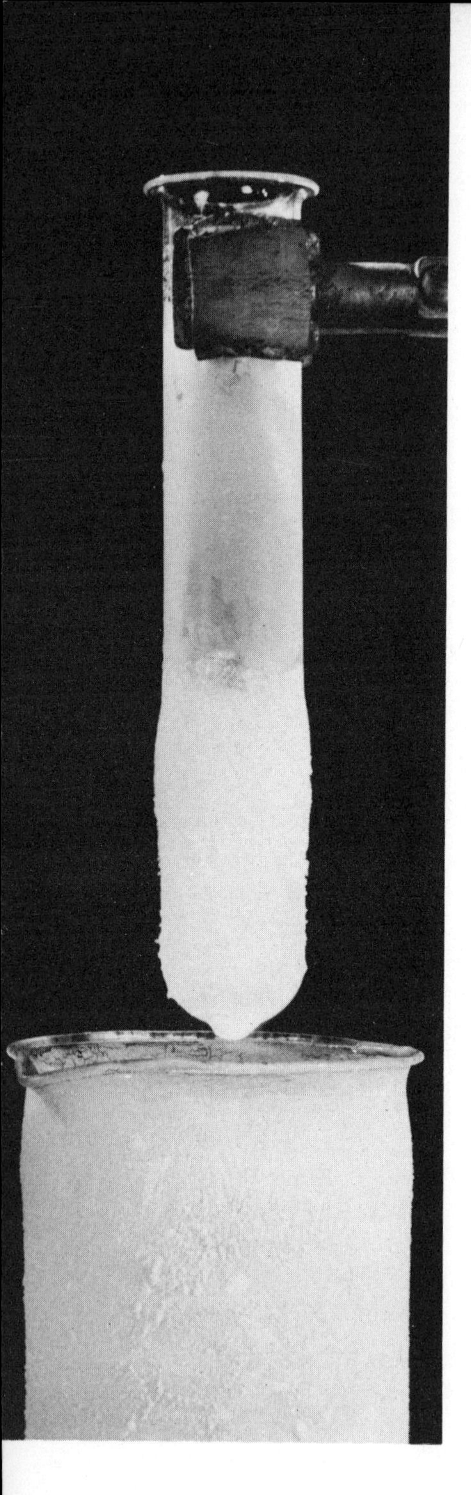

The cooling unit is made by filling a 250-ml beaker half full of denatured alcohol (*flammable!*) and adding small pieces of dry ice (*Caution: Handle dry ice with a cloth or tongs and keep your fingers out of the mixture*). Add the pieces slowly until violent bubbling subsides, then put in more to almost fill the beaker.

**To operate the apparatus,** open the clamp on the dropping funnel so that the acid falls on the sodium sulfate drop by drop. Reaction between the two chemicals—which may be speeded up by applying gentle heat intermittently under the flask—produces sulfur dioxide. This gas passes through the drying tube and down into the chilled test tube. There, cold of more than 70° below zero centigrade changes it into a liquid.

**Boiling sulfur dioxide makes ice.** When a few milliliters of liquid sulfur dioxide have collected, disconnect the apparatus and lift the test tube (holding it with a clamp) from the freezing mixture. (*Caution: Be sure the room is well ventilated, as sulfur dioxide is very irritating.*) The liquid will boil rapidly, absorbing enough heat to do so from the tube and the surrounding air. Just as on the freezing coils of your refrigerator, moisture from the air will condense on the tube, freeze there, and remain as long as the boiling continues.

When a quantity of liquid sulfur dioxide has collected, lift the test tube from the mixture in a well-ventilated room. The liquid will quickly evaporate, forming frost on the outside of the tube.

You can demonstrate the making of cold from chemicals by means of this improvised refrigerator. The container is a large fruit-juice can, wrapped with a towel for insulation. The cooling chemical is just a pound of photographer's hypo dissolved in water.

**Refrigerators operated by a gas flame,** or other source of heat, produce their cold by the evaporation of ammonia. The secret of changing the gas back into a liquid, without using compressors, lies in the fact that about 700 volumes of ammonia can squeeze itself into 1 volume of water at room temperature. The ammonia is first absorbed in water. Then it is driven out by heat, cooled by radiators, and finally squeezed into liquid form by its own pressure.

The mystifying gadgets that produce cold by dissolving chemicals in water are really no more wonderful than your ice box or refrigerator. When a solid goes into solution its molecules gain a greater freedom of motion. To do so they must absorb heat—just as ice must absorb heat to melt and water to evaporate. If a solid produces *heat* when it dissolves, this is due to a chemical reaction between solid and water that gives off more heat than is absorbed.

You can give a practical demonstration of cold from chemicals by cooling a bottle of soda on a picnic, in a household emergency, or as a party surprise. All you need is a large fruit-juice can, a turkish bath towel, 1 lb of common photographic hypo, water, and a couple of rubber bands. Fold the towel and wrap it around the sides and bottom of the can for insulation. Then pour in 1 qt of the coldest water you can get, dissolve the hypo in it by rapid stirring, and put in the bottle to be cooled. The temperature of the bottle should go down about 15°C. (The hypo solution need not be wasted. Bottle it and use it for preparing fixing baths.)

A mixture of 50 parts ammonium chloride and 50 parts potassium nitrate, dissolved in 160 parts water, will produce even greater cold. If the ingredients are originally at moderate room temperature, solution will cause a drop to below the freezing point of water.

Here is a list of typical mixtures with which you can produce low temperatures for your experiments.

**FREEZING MIXTURES**

| *Substance* | *Parts substance* | *Parts water* | *Initial temp. °C* | *Resulting temp. °C* |
|---|---|---|---|---|
| Ammonium nitrate | 100 | 94 | 20.0 | −4.0 |
| Ammonium nitrate | 60 | 100 | 13.6 | −13.6 |
| Ammonium chloride | 30 | 100 | 13.3 | −5.1 |
| Sodium thiosulfate (hypo) | 110 | 100 | 10.7 | −8.0 |
| Sodium chloride (salt) | 33 | ice, 100 | 0.0 | −21.3 |
| Mush of solid carbon dioxide lumps in alcohol | | | | −72.0 |

Trails made by alpha, beta, and gamma radiations from exploding atoms of radium can be seen in this easily made cloud chamber.

# Uranium and Atomic Energy

YOU can't make an atom bomb in your home laboratory—thank goodness!—but you can easily identify uranium, demonstrate the radioactivity of this unique element which makes the atom bomb possible, and build a cloud chamber in which you can track the alpha, beta, and gamma missiles of atomic disintegration.

Top member of the list of naturally-occurring elements—element number 92—uranium was discovered by the German chemist Klaproth in 1798 and named by him after the planet Uranus, then recently discovered by Herschel. For nearly a hundred years uranium remained in obscurity. Twice since, however, it has opened new epochs in scientific history.

Test ore suspected of containing uranium by placing it on screen on top of wrapped film. If true, ore will produce X-ray image.

The first was in 1896, when the French physicist Henri Becquerel discovered accidentally that uranium gave off rays which could affect photographic plates through black paper and even sheets of metal. Further investigation proved that these rays were the result of *exploding atoms*. Here was the first evidence that atoms were after all not really the ultimate building blocks of the universe!

The second was in 1939, when it was found that a rare variety of uranium, U-235, could be split artificially with the release of enough energy to split other atoms. From this discovery was born not only the atom bomb but all the peaceful developments of atomic energy that have since taken place. U-235 is made from ordinary uranium, U-238.

Strangely enough, uranium is as plentiful as tin, and more so than gold, silver, or mercury. Minute amounts may be found in nearly every type of rock and even in sea water. Because it is normally so thinly distributed, however, it is difficult to obtain. Places containing rich uranium ore are therefore exceedingly valuable.

**Uranium makes an X-ray picture.** If you think you have made a rich strike of uranium (or have bought a specimen of ore and wonder if you have been cheated) you can easily find out by means of a piece of photographic film wrapped in lighttight black paper plus a bit of wire screen or a paper clip.

Use a high-speed film and cut and wrap it in a totally dark room. Lay the package in a place where it won't be disturbed, making sure that the emulsion side of the film faces up. Place the screen or paper clip on it,

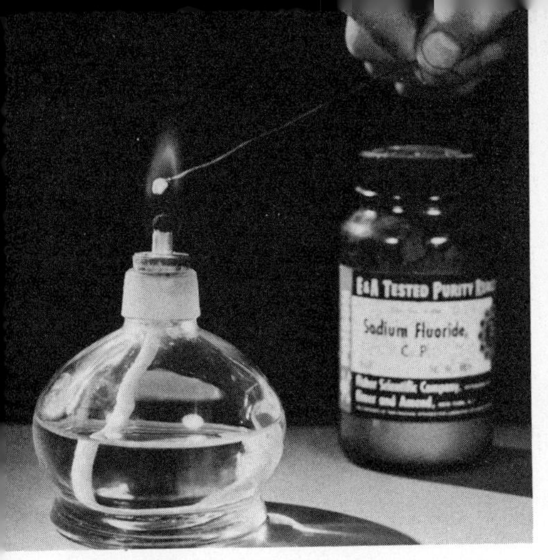

A quick test for uranium is to heat a speck of ore with sodium fluoride. Uranium will cause speck to glow under "black light."

and put the specimen of ore on top. If, on development of the film after 24 hours, you find a clear image of the screen or clip, you probably have a good sample of pitchblende. Carnotite and lower grades of uranium minerals may take from several days to several weeks to produce an image of equal intensity.

**Test for uranium under ultraviolet light.** A quick and almost positive test for uranium involves the glow or fluorescence of a uranium chemical under "black light." Although most uranium minerals will not fluoresce in their natural state, they will all do so when fused with sodium fluoride in a "bead test." As equipment for this test, you will need a few specks of finely ground ore, a 6-in. length of iron wire with a ¼-in. loop bent in one end, a bunsen, Fisher, or alcohol burner, some sodium fluoride, and a source of ultraviolet light. In the laboratory, the latter may be the bulb described on page 162, or it may be a special battery-operated UV light of a type made for use in prospecting.

To make the test, first heat the wire loop to redness, dip it in a little mound of the fluoride so that it picks up enough to fill the loop, and reheat it until the chemical melts to form a clear bead. Then touch the hot bead to a few grains of the ore and heat it once more until the ore has fused into it. When the bead has cooled, view it in a dark room under the ultraviolet light. If the sample contains even a trace of uranium, the bead will fluoresce to a bright yellow-green.

**Cloud chamber shows atomic trails.** Perhaps the most useful tool of the atomic-energy chemist and physicist is the cloud chamber, invented

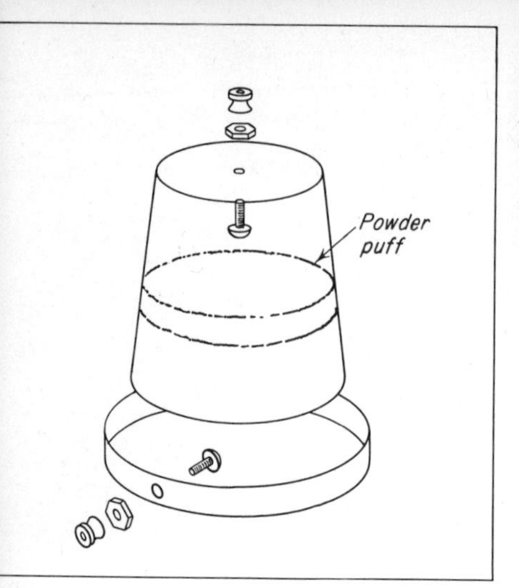


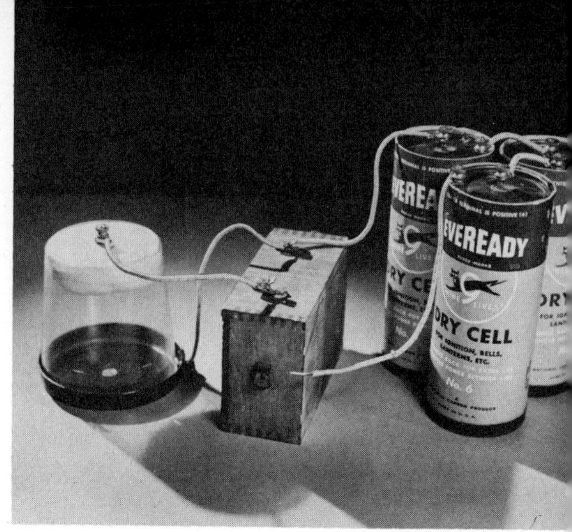


Construction of cloud chamber is shown at left, and wiring to the coil and battery above. The coil is used intermittently.

in 1911 by the British atom pioneer C. T. R. Wilson. The heart of this "showcase of atomic wonders" is an atmosphere supersaturated with the vapor of alcohol or other solvent. As atomic rays speed through this vapor, they cause molecules in their path to condense—thus leaving behind trails of visible droplets or "clouds."

In the original cloud chamber, supersaturation was produced by suddenly withdrawing a piston, thus cooling the vapor-laden air. This operation had to be repeated every time new trails were to be observed.

**The cloud chamber to be described** is a new type which has no moving parts and operates continuously. In it, alcohol vapor is caused to diffuse downward through a layer of air cooled by dry ice. At a point toward the bottom of the chamber, the vapor becomes cold enough to producc a state of supersaturation. To clear the atmosphere of ions and dust, and thus make new trails more distinct, provision is made to apply a high voltage intermittently between the top and bottom of the chamber.

Construction of the chamber is shown clearly in the diagram above. The chamber itself is an inverted, 1-pt plastic container. The hole for the binding post can be melted through with a nail heated to just below red heat. After the binding post has been fastened in place, attach a powder puff to the inside of the bottom with a few drops of household cement. For the base of the completed chamber use a metal can top, painted black on the inside with alcohol-resisting enamel, and provided on one side with another binding post. This top need not fit snugly.

High voltage is supplied by a Model-T Ford spark coil (obtainable

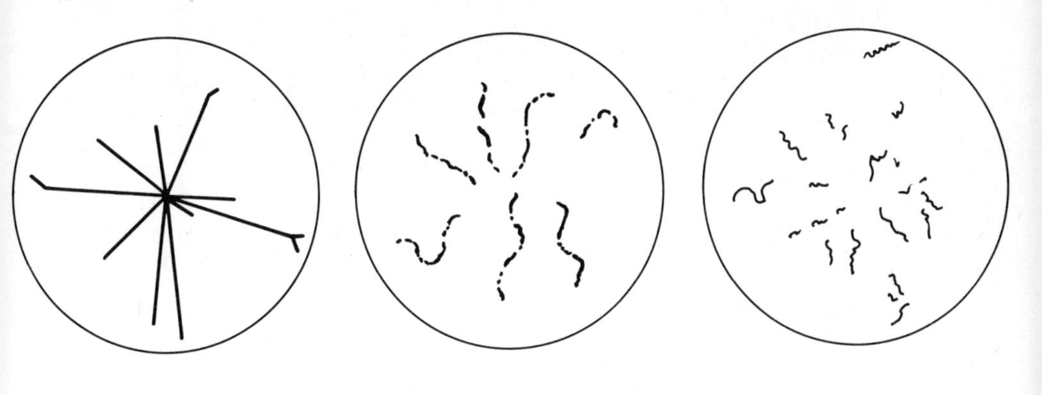

Alpha rays leave straight heavy cloud trails; beta rays, crooked and beady ones; short jiggles are produced by gamma radiation.

new from auto supply houses, if you haven't an old one) connected to two or three dry cells in series. For convenience, solder a binding post to each of the three terminals of the coil. Make connections as shown in the photo on the facing page.

A slab of dry ice (obtainable from an ice-cream distributor) as big as the base of the chamber will provide the cold. Lay this (*Caution: Don't handle dry ice with your bare hands*) on a cushion of cotton in a small pie tin to insulate and hold it. An excellent source of radiation is a speck of radium compound scraped from an old luminous watch or clock dial, attached to the tip of a thumbtack with a tiny dab of cement. Light for viewing can be supplied by a slide or movie projector or a strong flashlight.

**To operate your cloud chamber,** put the can cover on the dry ice, place the thumbtack, point up, on the middle of the cover, then place the plastic chamber on the cover—after having saturated the powder puff with denatured alcohol. For best seeing, view the chamber at about right angles to the direction of the light, as shown on page 135.

Within minutes, cloud trails will begin shooting from the radium. Straight heavy trails mark the path of alpha particles—atomic bullets with the mass and electric charge of helium nuclei. Crooked and beady trails are left by beta rays, high-speed electrons. Short jiggles are made by electrons knocked out of molecules by gamma rays, which hurtle through the chamber at the speed of light. Connection of the coil to the battery, when necessary, will clear the chamber of murkiness.

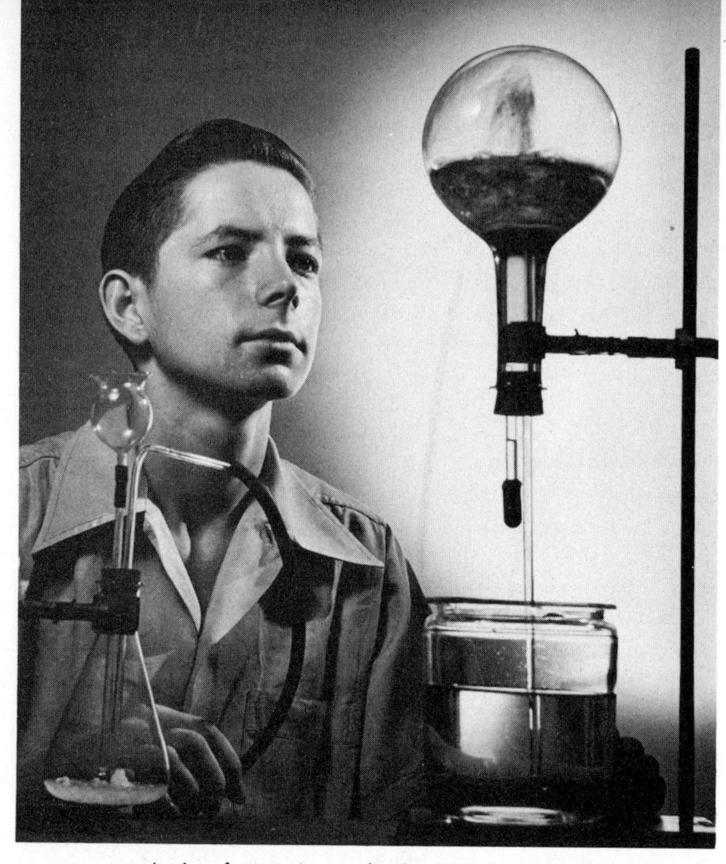

A pint of ammonia gas dissolves in a few drops of water in the flask, creating a partial vacuum that starts this fountain.

# The Chemistry of Solutions

WHENEVER you stir sugar into a glass of lemonade, you perform a feat of magic more wonderful than any on the stage. With a twist of the spoon, you make a solid vanish so completely that no microscope is powerful enough to detect it. Your taste will tell you, though, that the sugar is still present and in equal amount in every drop of the liquid. Let the water evaporate, and, lo! mixed with solids from the lemon juice, you will find your sugar back again.

All solutions involve this same magic. The dissolved substance, or "solute," disappears completely and uniformly in the dissolving substance, or "solvent." In a true solution, the solute will never separate from the solvent of its own accord. Yet remove the solvent by evaporation or other means, and the solute reappears.

Exactly what happens when a substance goes into solution, nobody knows. It is evident, however, that the solute breaks down into individual or small groups of molecules that dart about in every direction distributing themselves uniformly throughout the solvent.

**See molecules in motion.** How this motion begins can be beautifully demonstrated by dropping a few grains of a water-soluble dye into a tall glass or cylinder of water. At first the dye falls straight down, pulled by gravity; then the grains swirl with the water currents. Finally molecular movements, too small to see individually, spread the dye uniformly throughout the water.

**Dissolving produces heat and cold.** Heat is required to free molecules so they can move about. This heat is extracted from the solvent and it is cooled. At the same time some substances react chemically with the solvent and produce heat. If the cooling and heating effects balance, the temperature won't change. If they don't, the solution may end up considerably hotter or colder than the original ingredients.

You may easily demonstrate this paradox of solution with sodium hydroxide (ordinary lye will do) and ammonium nitrate. Stir the chemicals into separate glasses of water at room temperature and test with a thermometer. The temperature of the lye solution will be near the boiling point, while the other will be at freezing or below.

Chemical heating pads, freezing solutions, and coolers for food and drink all depend upon a similar heat imbalance produced when certain substances dissolve.

Dropped in water, dye first falls straight down. Then water-swirls and molecular activity carry it until it is distributed uniformly.

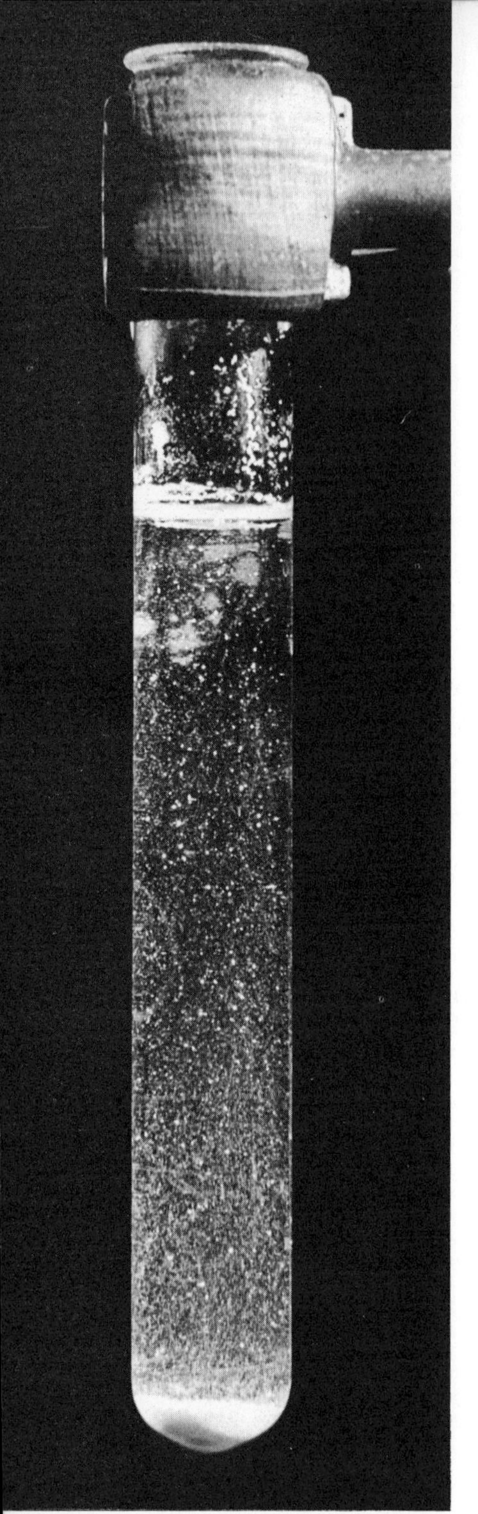

**Temperature governs solubility.** Temperature also determines the amount of a given substance that will dissolve in a particular solvent. Except for a few chemical compounds such as some calcium salts and certain organic acids, solubility increases with a rise in temperature—not much in some cases, enormously in others. You can dissolve about 36 g of common salt or 31 g of saltpeter in 100 ml of water at room temperature, for example. At the boiling point, you can dissolve only 3 g more of salt, but 215 g more of saltpeter!

**A shower of lead iodide.** Most solutions that are saturated at high temperature must be kept at that temperature or the solute will crystallize out. For a fascinating demonstration of this, dissolve as much lead iodide as you can in a test tube of boiling water, stirring for several minutes. Then support the tube upright and let the solution cool. Tiny golden crystals will begin to form all through the solution, growing before your eyes and finally falling slowly in a beautiful golden shower.

**Supersaturated solutions are unstable.** A *saturated* solution is generally defined as one containing as much dissolved solute as it can hold. This is not exactly true, for with care, and with certain chemicals, you can make solutions which are more than saturated. Such "supersaturated" solutions, however, are unstable and can readily be made to part with their excess solute.

When a hot saturated solution of lead iodide cools, tiny crystals will form spontaneously and fall in a glittering golden shower.

Supersaturated solution may be liquid one moment and solid the next. A solution of hypo (left) solidifies when stirred (right).

With ordinary photographers' hypo (sodium thiosulfate), you can demonstrate this. Half fill a small beaker with hypo crystals and place the beaker over a small flame. The crystals will soon form a liquid as they dissolve in their own water of crystallization. If the beaker is clean and if there are no foreign particles or undissolved crystals in the solution, the solution may be cooled without having crystals form. Add a single crystal of hypo, however, or even scratch the inside of the cooled beaker with a stirring rod, and crystallization begins, causing the whole solution to solidify.

A more accurate definition of a saturated solution therefore is that it is a solution that will not gain or lose in concentration if a little undissolved solute is added. In your practical laboratory work, you may consider a solution saturated if, after a reasonable amount of stirring, some of the solute settles to the bottom.

**Because it dissolves so many things,** water is called a "universal solvent." When a substance is said to be "soluble," the solvent is assumed to be water unless another is mentioned. There are dozens of other solvents. Alcohol, ether, acetone, carbon tetrachloride, are a few of the commoner ones.

**Other kinds of solutions.** When we speak of solutions, we generally think of solutions of a solid in a liquid. There can be solutions, however, of liquids in liquids (alcohol in water, for example), gases in liquids (household ammonia), and even solids in solids. Solder, brass, and many other alloys are so united chemically that they are often spoken of as "solid solutions."

Soda water is a solution of carbon dioxide gas in water. It bubbles as it warms up to room temperature because, unlike solutions of solids in liquids, the saturation point of gases in liquids decreases as the temperature rises. Boiling will eliminate the gas entirely.

**Amazing solubility of ammonia explains a fountain.** At room temperature, water can dissolve a little less than its own volume of carbon dioxide. At the same temperature it can, however, dissolve the amazing amount of 442 times its own volume of hydrogen chloride or 700 times its volume of ammonia gas. The extreme solubility of these two gases can be used in performing the dramatic fountain demonstration shown at the beginning of this unit.

Ammonia is the easier of the two to make and use. Set up a 500-ml round-bottom flask as indicated. It is important to use a round-bottom flask, for one with a flat bottom might be crushed by the air pressure. Pass through one hole of a 2-hole stopper a long, straight glass tube drawn out to a jet at its upper end, and let the lower end dip into a vessel of water to which a little phenolphthalein solution has been added. Then pass a medicine dropper filled with water through the other hole.

**You may make ammonia** in a generator like that shown in the photo by adding 28 per cent ammonium hydroxide through the thistle tube to sodium hydroxide pellets on the bottom. Remove the stopper from the round-bottom flask and lead ammonia gas into it from the generating flask through a rubber tube. Being lighter than air, it will displace the air downward. When an ammonia smell issues from the mouth, remove the rubber tube, and stopper the flask tight.

Now squeeze water from the medicine dropper into the inverted flask. Ammonia will dissolve, and a partial vacuum will be created, causing water to rise from the lower vessel and squirt out the jet. Since ammonia is an alkali when dissolved, it turns the phenolphthalein in the water pink.

As more water enters the flask, more and more ammonia dissolves, and the vacuum becomes increasingly greater. Finally the water shoots right to the top of the flask and continues to do so until the flask is almost full.

Ferric chloride solution, added slowly to boiling water, yields colloidal particles of ferric hydroxide that stay in suspension.

# Colloids—a Special State of Matter

NEVER met a colloid? Then you have never eaten a piece of bread, washed your hands with soap, mixed paint, or performed any of a thousand everyday things that bring you face to face with a fascinating branch of chemistry.

In a paper published in 1861, Thomas Graham, a Scottish chemist, observed that many chemical substances fell into two classifications: those which, dissolved in water, would pass readily through an animal membrane, and those which would not. The former were substances which crystallize from their solutions, like salt or sugar; the latter—substances such as gum arabic, albumin, and gelatin—were never known as crystals. Graham called the first class "crystalloids," and the second "colloids," from the Greek word *kolla,* meaning glue.

Today, however, chemists believe that almost any substance may act either as a crystalloid or colloid, depending upon the size of its particles. A colloid, they hold, is an intermediate subdivision of matter, too big to dissolve completely, yet too small to be seen with even the most powerful optical microscope.

**Colloids scatter light.** Although colloidal particles themselves are invisible, you may demonstrate their presence simply. Fill one bottle with water containing a little rubbing alcohol, and another with water containing a few grains of salt. Both liquids look equally clear, but turn out the ordinary room light and shine a strong pencil of light from a flashlight or spotlight through the bottles at a right angle to your line of vision. The difference is amazing. Passing through the salt solution, the light is practically invisible; in the alcohol and water, the beam shines out strongly. Why? The salt particles are so small that they cannot reflect the tiny light waves, but the colloidal particles formed by mixing alcohol with water are large enough to scatter the beam.

**Anything may be a colloid.** Under suitable conditions it is possible to produce almost any substance in colloidal form. This may be done by one of several methods. The particles may be built up, or condensed, by the reaction of atoms, molecules, or ions; they may be broken down, or dispersed, by the disintegration of larger particles.

**How to make colloidal iron.** You can demonstrate the first method in your own lab. Ferric hydroxide can be easily condensed into colloidal particles by adding a concentrated solution of ferric chloride to boiling water. The chloride is rapidly changed by the water into particles of ferric hydroxide so minute that they may remain suspended for years.

To work this transformation, add several drops of ferric chloride solution to a beaker of boiling water, set swirling by stirring. Almost instantly the pale yellow of the ferric chloride changes to the deeper reddish-brown of ferric hydroxide.

Enough of this colloidal suspension for further experiments may be made by adding 50 drops of a concentrated solution of ferric chloride, freshly prepared, to 500 ml of water that is just barely boiling. Add water occasionally and boil until the suspension becomes clear and a deep reddish-brown. (By boiling too vigorously, or by adding the chloride solution too fast, you may cause the ferric oxide to be precipitated. If that happens, you must begin again.)

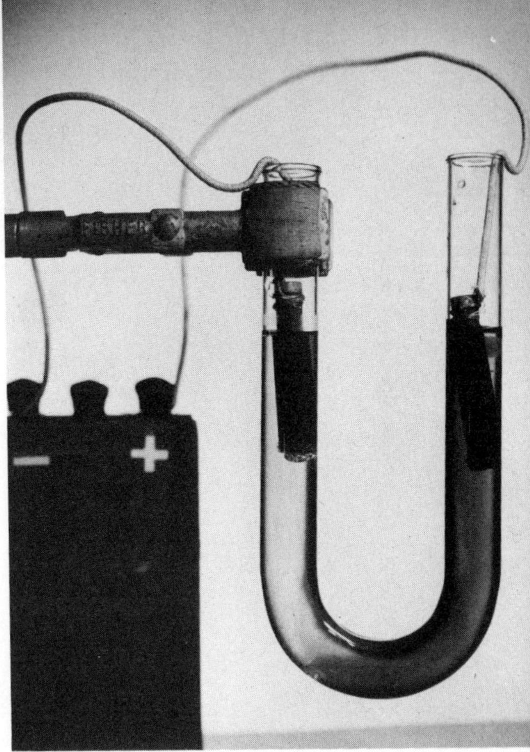

Apparatus for dialysis is at left. That colloidal suspensions have an electrical charge can be shown as above. Here particles are repelled by the negative pole and attracted by the positive.

**Dialysis separates iron from acid.** When ferric chloride is thus changed into ferric hydroxide, hydrochloric acid is also formed. To remove it from the suspension, you must use Graham's famous method of separation, which he called "dialysis." Both acid and colloid will pass through the pores of an ordinary filter, but a membrane with much finer pores will hold back the ferric hydroxide particles.

Graham used pig bladders and parchment paper, but today collodion films and ordinary nonwaterproofed cellophane serve more effectively. Cut a disk of the latter about eight or ten inches in diameter, shape it into a bag, and tie the end tightly around the stem of a thistle tube or small funnel. Half fill this bag with the suspension, and hang it in a large beaker of water. After a few minutes, test the water surrounding the bag with blue litmus paper. The paper will turn pink, indicating that the water has become acid. Continue the dialysis for several hours, changing the outside water frequently.

**Particles are electrified.** One reason that colloids stay in suspension against the pull of gravity is because the molecular movements of the

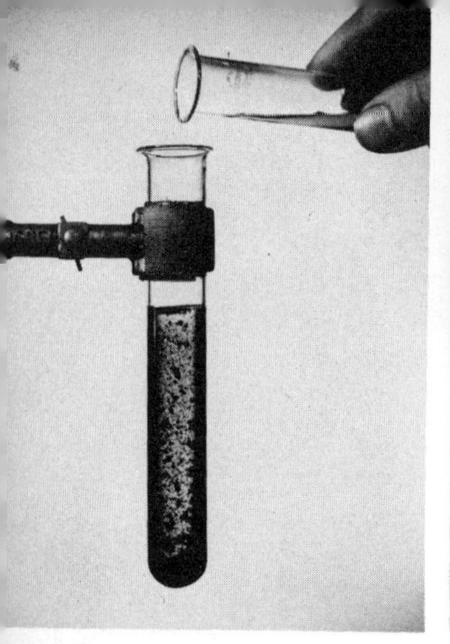

Negative and positive colloids neutralize each other and fall out (left). Colloidal blue goes through filter; coarser blue is held back (center). Peptization lets coarse blue go through.

sustaining liquid bounce the small particles back and forth. Another is that most colloids repel each other because each contains an electrical charge. You can demonstrate the latter property by attaching a 45-volt or 90-volt radio B battery to carbons taken from two flashlight cells and inserting the carbons in a glass U-tube filled with a suspension.

A suspension of colloidal prussian blue responds strikingly to this test. You can make this suspension by adding, with gentle stirring, just 2 drops of a saturated solution of ferric chloride to 100 ml of water which contains 12 drops of saturated potassium ferrocyanide solution. Immediately the previously water-white solution turns a deep blue. Dialyze this blue suspension as you did the ferric hydroxide. Then pour it into your U-tube and apply voltage. After a few minutes, you will notice the blue coloring is being pushed from the negative electrode and is concentrating at the positive.

Prussian blue in suspension form is negatively charged; ferric hydroxide, on the other hand, is positively charged, as can be proved by repeating the experiment with the latter chemical. The movement here, however, may not be so pronounced, for this substance tends to precipitate out when it concentrates at the negative pole. Many dyes are colloidal and are selectively attracted to different types of chemical fibers largely because of an electrical charge.

**Opposite charges break a suspension.** As you might guess, the particles in negatively and positively charged suspensions, when added together in the proper proportions, are attracted to each other. Therefore, the charges are neutralized, and the particles often fall out of suspension. You can verify this by adding ferric hydroxide suspension to some suspension of prussian blue. Large particles will form and fall to the bottom.

Acids, bases, or salts added in sufficient quantity to an electrically charged suspension also will discharge it and cause precipitation. In city water systems, this is the principle involved when aluminum or iron sulfate is added to clarify muddy water.

**Two forms of one substance.** Sometimes substances may be prepared, in either colloidal or grosser form, by controlling the concentration of the substances. If you pour prussian blue suspension into an ordinary filter, the blue will pass through, leaving no solid matter and hardly coloring the filter. But make a new batch of this blue, this time adding 10 drops of ferric chloride solution and 20 drops of potassium ferrocyanide to 100 ml of water, pour it into a filter—and note the difference. Although the first milliliter or two of the filtrate may be pale blue, the remainder will be colorless and a precipitate of the blue pigment will remain in the filter. Chemically, the two prussian blues are almost identical; physically, the first is colloidal while the second is not.

One method of breaking down large particles to make them of colloidal size is called "peptization." You may demonstrate this process by pouring a dilute solution of oxalic acid through the precipitate of gross prussian blue on your filter. Immediately the blue begins to pass through the pores of the filter. If the filtrate is subsequently dialyzed to remove the oxalic acid, the blue will remain in colloidal form.

Colloidal prussian blue was once an important writing ink, but aniline dyes are now more commonly used.

**Colloidal particles may often be produced** by diluting a solution containing some solid with another liquid in which the solid will not dissolve. For instance, by pouring an alcoholic solution of sulfur into water, a milky suspension of sulfur can be formed. The whitening of newly applied shellac by drops of water is another example, the water diluting the alcohol in which the shellac resin is dissolved. As the resin will not dissolve in the water, it is precipitated as colloidal particles that disperse light and look white.

Sodium oleate is produced and an emulsion forms when you add oleic acid to an oil, sodium hydroxide to water, and mix the two.

# When Oil and Water Mix

EVERY time we drink milk, eat eggs, or apply one of the new oil paints that can be thinned with water, we are dealing with a contradiction. "Oil and water won't mix," says the old adage. Yet in milk, eggs, emulsion paints, and in hundreds of other products that combine only with watery substances, nature on the one hand and the chemist on the other have evoked chemical trickery to disprove the old rule and perform the seemingly impossible.

Emulsions are the means by which it is done. Without these strange combinations of oily and watery substances, science, medicine, and industry would face serious handicaps; in fact, life could hardly go on. Fats and oil-like hydrocarbons are needed in the growth of animal and plant cells. Yet the main substances of these living cells are watery.

Unable to dissolve the oils in water, nature uses emulsification. An almost perfect food is provided in milk, an emulsion of butter-fat globules in a water solution of lactose, mineral salts, and protein.

**Emulsions work modern wonders.** By means of emulsions, unpleasant castor and cod-liver oils can be made almost palatable, and foul-smelling oily substances are made odorless. Paints can be formulated to dry with a tough pigment-in-oil surface, although they are easily thinned with plain water. Modern agricultural sprays, salad dressings, cosmetics, lotions, drug preparations, and floor, furniture, and shoe polishes often owe their effectiveness to the know-how of the emulsion chemist.

**What constitutes an emulsion?** Fundamentally, an emulsion is a dispersion or scattering of tiny droplets of one liquid in another with which it will not mix. An emulsion can usually be produced by shaking or stirring the two liquids together. But unfortunately, an emulsion formed in this manner from two pure liquids quickly breaks up. Surface tension causes the globules to unite rapidly, and the drops then either rise or fall, separating the liquids into two layers.

If you shake together in a test tube a little water and an equal amount of benzene you can see how this process works. After a few shakes, the two water-clear liquids will become a single milky liquid. (The liquid seems uniform because the droplets are too small to be seen with the naked eye, and it is white because the droplets, acting like lenses, scatter the light.) Quickly, however, visible drops appear. These become larger and larger, the whiteness vanishes, and in a matter of seconds the liquids will have separated.

**Agents keep emulsions from breaking.** To keep such an emulsion from breaking up requires all the magic of the chemist, for a stable emulsion can never be formed from two pure liquids alone. He must add a small quantity of a third substance, known as an "emulsifying agent." Since there are hundreds of possible substances that can serve as agents, each having its own special merits, the chemist must know how to select the best one for the emulsion he requires.

Plain soap (sodium stearate or sodium oleate) is a common emulsifying agent. Repeat the previous water-benzene experiment after first dissolving a few shavings of toilet soap in the water. In this case, intermittent shaking, with a pause of 15 seconds after every few shakes, often will produce emulsification faster than continuous shaking. After shaking

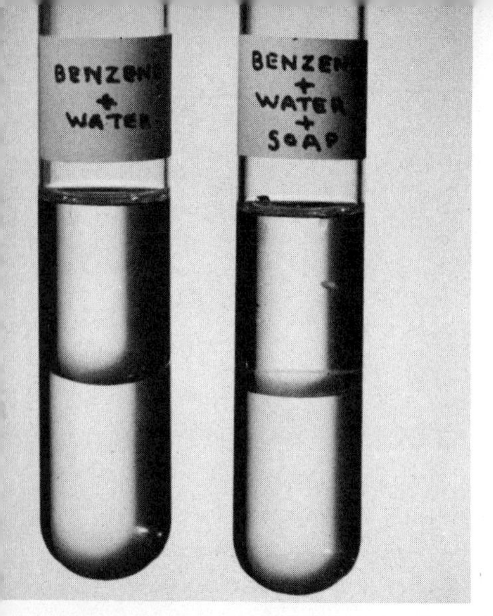

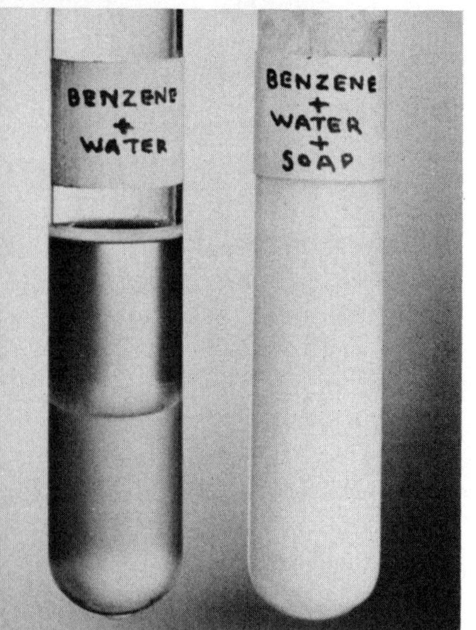


Shaken vigorously, water and benzene can be mixed, but a substance known as an "emulsifying agent" is needed to keep them from separating again. To prove this, prepare test tubes as shown at top. Now shake both tubes, then let them stand. In one tube, the liquids separate. In the other, soap keeps them emulsified.

intermittently for 5 minutes, your emulsion looks the same as the one produced with only water and benzene.

But this time the oil and water will separate only slightly, if at all. However, after long standing, this emulsion may appear thicker and whiter at the top. The effect is called "creaming" because of its similarity to cream rising on milk. It does not indicate "breaking," the permanent separation of the oil from the water. When creaming occurs, a slight shaking will make the emulsion uniform again.

**Soap as an emulsifying agent.** Using soap as the emulsifying agent, you can similarly emulsify kerosene or medicinal mineral oil with water. Instead of adding ordinary soap, you can vary the procedure by making a special soap at the same time you are producing the emulsion. To do this, add a few grains of sodium hydroxide to 10 ml of water in one test tube and a little oleic acid to a similar amount of oil or benzene in another test tube. If you then pour the contents of the latter tube into the former, the soap sodium oleate is formed spontaneously wherever the oil and water touch each other. In this case an emulsion forms almost of itself. It can be stabilized by a minimum of shaking.

Incidentally, soap helps clean your skin and your clothes partly by enabling water to mix with oil and grease, thus forming an emulsion which may easily be rinsed away.

**The theory behind emulsions.** Besides sodium and potassium soaps, emulsifying agents include such substances as egg yolk, gelatin, gum arabic, gum acacia, sulfonated oils, aluminum and zinc stearates, and such finely powdered substances as mustard, clay, silica, and carbon black.

The "why" of emulsifying agents is complex and imperfectly understood. In most cases they reduce the tension between the two liquids. Sometimes they set up an actual physical barrier that prevents the globules from joining.

In the case of soaps and synthetic organic compounds, the protective effect is believed to be due to a unique molecular structure. One end of each molecule of such substances is soluble in oil and the other in water. When placed in a mixture of oil and water, these molecules tend to orient themselves so that their oil-soluble ends are turned toward the oil and their water-soluble ends toward the water. By this action, each suspended globule is finally encased in a sheath of molecules that at once bind it to the surrounding liquid and at the same time prevent it from coalescing with other globules of its own kind.

All emulsions consist of two "phases": an external or continuous phase, and an internal phase which consists of millions of separate tiny droplets. Depending upon the emulsifying agent and sometimes upon the conditions of mixing, an emulsion may consist of oil in water or of water in oil. In technical literature, an oil-in-water emulsion is abbreviated O/W, while a water-in-oil emulsion is W/O.

Milk, mayonnaise, the benzene-in-water emulsions you have just made, and most emulsion paints are of the oil-in-water type. Butter, oleomargarine, and some lubricating greases are water in oil. Often the two types of emulsion may look alike and tests must be made to determine which is which.

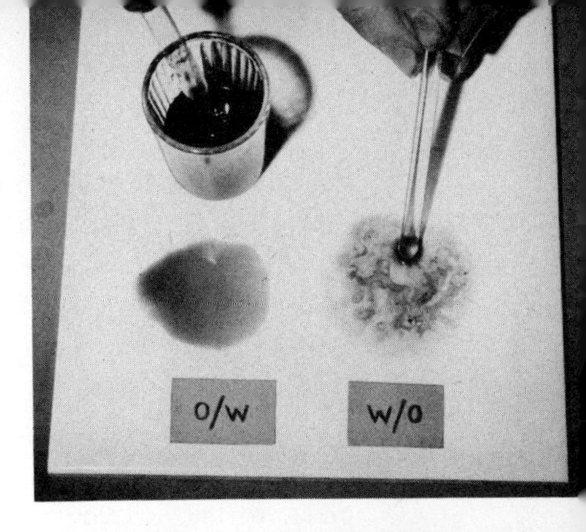


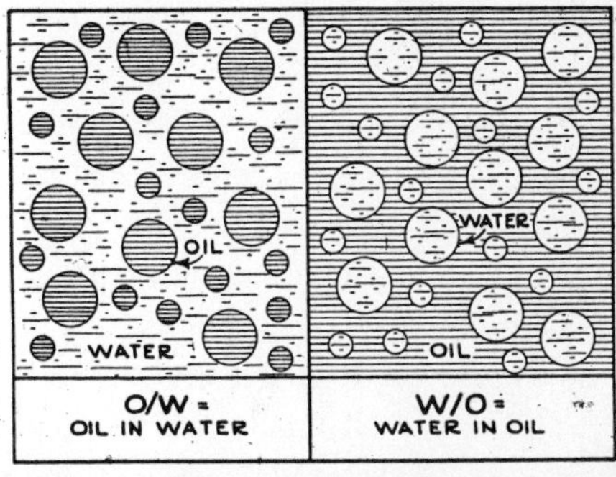

Emulsions consist either of oil in water (O/W) or water in oil (W/O). By adding a water-soluble dye to a sample of each, as at top, you can find out which is which. If the dye blends, the emulsion is O/W; if it does not, it is W/O. The bottom photo shows how water thins (left) and oil thickens an O/W emulsion.

**Oil in water or water in oil?** One test is to add to a sample of the emulsion a few drops of a dye that is soluble only in oil or only in water. For example, add a water-soluble dye to a little of your benzene-water emulsion and stir it gently. The sample will be colored uniformly, indicating that the continuous or external phase of the emulsion is water. If the emulsion were water in oil, the water-soluble dye could not color it continuously, but would dye it only in spots. (You can make a W/O emulsion for this comparison by adding a little calcium or magnesium oleate to the benzene, instead of soap to the water, before emulsification.)

**To thin or thicken any emulsion,** you must first know its type, for if the wrong phase is added you will produce the opposite effect from what you wish. To thin a given emulsion, you must add more of the external phase. To thicken it, you must add more of the internal phase. You can show this with your benzene-in-water emulsion. Add water to it and it flows more freely. Add benzene, with additional shaking, and it finally will become so thick that it cannot be poured.

Oddly enough, the internal phase of an emulsion may have considerably greater volume than the external. Mayonnaise is such an emulsion, of the O/W type, in which a large volume of oil is contained inside a volume of water that is comparatively small.

**Making a sample of palatable mayonnaise** affords a good demonstration of some of the problems of commercial emulsion making. In this foodstuff, salad oil (either cottonseed or corn oils are suitable) forms the internal phase, vinegar or lemon juice the external phase, while egg yolk, powdered mustard, and paprika jointly make up the emulsifying agent. Salt must be added as seasoning, but it tends to destroy, rather than help, the emulsification.

Put an egg yolk, ½ teaspoon salt, ¼ teaspoon dry mustard, and ⅛ teaspoon paprika into a small bowl and beat thoroughly. Add 1 tablespoon of vinegar and beat again. Very gradually, beat in 8 oz (1 cup) of salad oil, adding only ½ teaspoon at a time for the first 2 oz. Then add ½ oz at a time. By the time you have added all the oil, you will have produced a first-rate mayonnaise.

You can easily measure the acidity or basicity of a solution by comparing it with a standard solution in a test called "titration."

# How Strong Is That Acid?

NEUTRALIZATION is one of the most important and frequently used operations of chemical testing and manufacturing. By mastering the technique, you can determine how much acid or how much base a solution of unknown strength contains. By combining acids and bases accurately, you can also form dozens of new compounds for experiments.

To demonstrate the principle of neutralization, all you need are a few grains of sodium hydroxide, a little hydrochloric acid, and a few drops of phenolphthalein solution.

**Phenolphthalein solution,** made by dissolving 1 g of the chemical in 50 ml of alcohol and then adding 50 ml of water, is used as an "indi-

When hydrochloric acid neutralizes the sodium hydroxide solution at left, and the water is evaporated, the result is table salt.

cator" of basicity. Add just 2 drops of the solution to a base which you have made by dissolving about a gram of sodium hydroxide in 50 ml of water, contained in a small glass. A mere 2 drops turns the hydroxide solution a bright pink. As long as the color remains, you can be sure that the solution is still basic.

For your acid, mix about 3 ml of concentrated hydrochloric acid thoroughly in 12 ml of water. Add this acid very slowly to the sodium hydroxide solution, stirring the latter constantly with a glass rod. When the pink color begins to lighten, add further acid only one drop at a time, using a medicine dropper. You will notice that at this stage parts of the solution will clear and then turn pink again. Continue even more slowly than before, pausing and stirring between drops. Finally, the addition of one last drop of acid will cause the entire solution to become colorless.

**Two deadly poisons make common salt.** At this point, called technically the "end point," acid and base have exactly neutralized each other. What remains in solution is a salt, which is neither acid nor basic. If you evaporate the water and analyze the compound that remains, you will find it to be sodium chloride. Thus, two substances which by themselves are poisonous have been transformed, by the simple magic of neutralization, into common table salt!

**Many other salts may be made just as easily.** Sodium hydroxide, for instance, neutralized by dilute sulfuric acid produces sodium sulfate; by nitric acid, sodium nitrate; by acetic acid, sodium acetate, and so on.

In neutralizing, 40 g of sodium hydroxide mixed with 36.5 g of hydrochloric acid form 58.5 g of table salt and 18 g of water.

Substitute potassium, barium, strontium, or other soluble hydroxide for the sodium, and your salt becomes a compound of one of the respective metals. Calcium hydroxide, cheapest of the bases, can be neutralized in the same way, but the procedure takes longer because this chemical is only sparingly soluble. As soon as the dissolved portion becomes neutralized, more of the chemical goes into solution.

Insoluble hydroxides, such as those of iron and copper, also react with acids to form salts. The end point, however, cannot be determined by the technique mentioned.

**To understand how neutralization works,** we must know what happens to acids and bases when they are dissolved. According to modern chemistry, an acid is a substance which in water solution dissociates or breaks up, producing as one component *hydrogen ions*—positively charged atoms of hydrogen. These hydrogen ions make solutions of acids taste sour. They turn blue litmus red.

Bases—metallic hydroxides, or combinations of a metal with oxygen and hydrogen—produce negatively charged *hydroxyl ions* when they dissociate in water solution. These are what make bases taste bitter and feel soapy. They turn red litmus blue.

Mix water solutions of acids and bases and what happens? Positive hydrogen ions unite with negative hydroxyl ions and form $H_2O$—plain water. Evaporate the water from the solution, and the remaining nonmetallic element of the acid and the metal of the base combine, thus forming a salt.

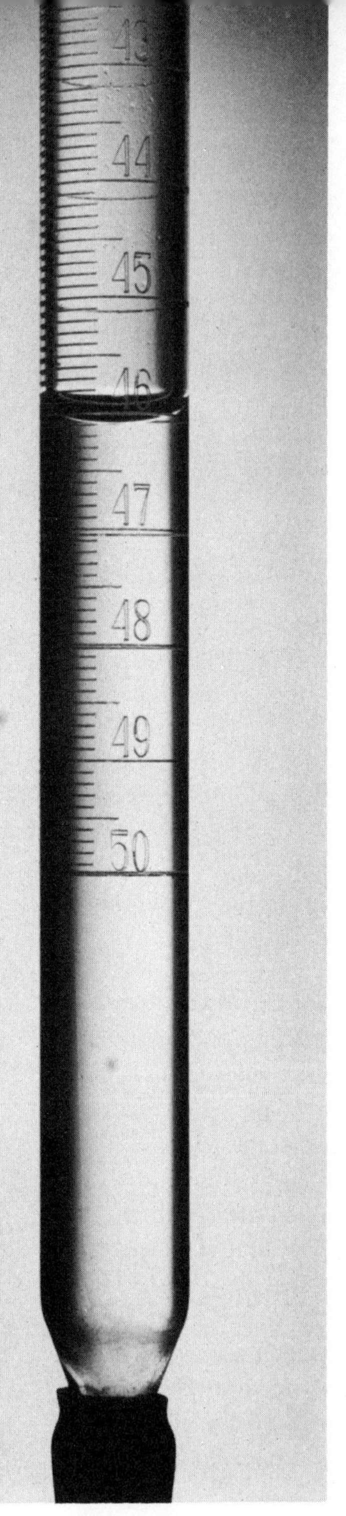


**Titration measures strength of acid or base.** Because acids and bases always unite in exact proportions, the process of neutralization provides a reliable and easy method of finding out the strength of solutions of these compounds. The measurement procedure is known as "titration." It is one of the most important in the chemical laboratory. The procedure involves running a solution of acid or base, of standard strength, into a solution of unknown strength until neutralization occurs. The strength of the unknown then is readily calculated by comparing the volume of standard-strength solution needed to neutralize a given volume of the unknown one.

A long, accurately graduated tube, known as a "burette," is used for measuring the exact amount of the standard solution needed for neutralization. Commercial laboratories generally provide a special stand that holds two burettes, one for acid and one for base, but a single one mounted by means of a clamp on a ring stand is fine for the beginner. A 50-ml burette, to be used with a pinchcock, costs less than $2.

**Standard solutions** employed in the process of titration are called "normal" solutions, generally abbreviated as *N*. Solutions which are multiples or decimal fractions of normal, such as 5*N*, 0.1*N*, 0.5*N*, and so on, also are used. Normal solutions are so adjusted that equal volumes of normal solutions of any acid and any base will exactly neutralize each

In reading a burette, keep your eye at the level of the lowest part of the curve of the liquid. Notice that the graduations read from the top downward.

**TABLE OF WEIGHTS**
**OF ACIDS AND BASES IN STANDARD SOLUTIONS**

| | Formula | Molecular Weight | Grams Per Liter Normal Sol. | Grams Per Liter 0.5 N Sol. | Grams Per Liter 0.1 N Sol. |
|---|---|---|---|---|---|
| *ACID* | | | | | |
| Hydrochloric | $HCl$ | 36.5 | 36.5 | 18.25 | 3.65 |
| Sulphuric | $H_2SO_4$ | 98 | 49 | 24.5 | 4.9 |
| Nitric | $HNO_3$ | 63.02 | 63.02 | 31.51 | 6.3 |
| Phosphoric | $H_3PO_4$ | 98.1 | 32.7 | 16.35 | 3.27 |
| Acetic | $HC_2H_3O_2$ | 60.03 | 60.03 | 30.01 | 6.00 |
| *BASE* | | | | | |
| Sodium hydroxide | $NaOH$ | 40 | 40 | 20 | 4.00 |
| Potassium hydroxide | $KOH$ | 56.11 | 56.11 | 28.05 | 5.61 |
| Barium hydroxide | $Ba(OH)_2$ | 171 | 85.5 | 42.75 | 8.55 |

other. For instance, a normal solution of acid contains 1 g of hydrogen ions per liter and a normal solution of base, 17 g of hydroxyl ions per liter.

**Hydrochloric acid** generally is used for making standard acid solutions. However, since it really is a water solution of a gas varying considerably in strength, we might get greater accuracy by using sulfuric acid. Dissolve 26.6 ml of concentrated sulfuric acid by pouring it slowly into 4 times its volume of water; then dilute to exactly one liter. This will result in a solution very close to normal. Solutions of 0.5*N* and 0.1*N* may likewise be made by using 13.3 ml and 2.66 ml respectively.

**Sodium hydroxide** is the base generally used. If 42.5 g are dissolved and diluted to a liter, you should have an approximately normal solution of base.

**How to test vinegar.** Equipped with these standard solutions and a burette, you'll have the means of determining the strength of solutions of most common acids and bases. Suppose, for instance, that you want to check the acetic acid strength of a particular brand of vinegar. (It would be best to use white vinegar so as to avoid any color confusion.)

Put a measured amount of vinegar, say 20 ml, into a small Erlenmeyer flask or beaker and add 2 drops of phenolphthalein solution. After filling the burette with normal sodium hydroxide solution, open the pinchcock slightly, allowing the excess base to run off into a clean beaker until the surface is exactly at the zero line. Notice that the numbers on the burette read from the top downward. Always sight directly through the burette in observing the volume of solution it holds, keeping your eye at the level of the lowest part of the curve of the liquid surface, which is known as the "meniscus."

**Technique of titration.** Now place the flask or beaker of vinegar under the burette and let the basic solution run slowly into it. Swirl the flask or beaker constantly or agitate the solution with a stirring rod. The mixture eventually will turn partly pink, but the color will disappear with further swirling or stirring. From then on, let out the hydroxide drop by drop. When, with the addition of 1 drop, the solution remains pink, the end point is reached and the titration finished.

Finally, compare the volume of vinegar used with the volume of base required to neutralize it. Suppose you needed 10 ml of base to neutralize the 20 ml of vinegar. Obviously, the vinegar is just half as strong as the base, or 0.5 normal. By referring to the table on page 159 you see that a 0.5 normal solution of acetic acid contains 30 g per liter. Since water weighs 1,000 g per liter, this proportionate weight of acid means your vinegar is a 3 per cent solution.

By reversing the procedure—putting a measured quantity of base into the flask and an acid solution into the burette—you can determine the strength of base solutions. In this case the solution is pink at the start and clears at the end point.

By coating half of a jar with fluorescent paint and turning it over a Purple-X bulb, you will learn how fluorescent lamps work.

# Sleuthing with Invisible Light

NO matter how good your eyes are, you are really part-blind. But don't worry, everybody else is too. It is this defect in human vision that makes possible the gleaming new colors you see on magazine covers, road signs, and teen-age clothing. Scientists use it not only to test ores but also to find out whether mice are susceptible to cancer.

And you can use it in your own lab for a variety of experiments—from seeing how clean your wife or sister washes the dishes to finding out whether Aunt Flora has false teeth.

**Fluorescent materials borrow invisible light.** Our eyes are sensitive to only part of the light that hits them: they are blind to wave energy at the two ends of the spectrum—out beyond red at one end and violet at the other. But there are a number of substances that can borrow invisible light from the violet end—the ultraviolet—and turn it into the kind of light we can see. A mineral called "fluorspar" was one of the first materials found to have this property, so it is called "fluorescence." Some vitamins, natural oils, minerals, even your teeth, are fluorescent. Placed under pure ultraviolet light, a fluorescent object gleams in color when everything else is black.

Secret messages written with fluorescent inks glow in this "black light." So do invisible fingerprints or dirt on a plate when dusted with fluorescent powder. And because different substances fluoresce in different colors, things that look alike are found to be quite different when studied under ultraviolet light.

**Daylight-fluorescent printing inks,** paints, and dyes achieve their unique brightness in the same way. A fluorescent red ink, seen in daylight, not only reflects practically all the red in the light that hits it, but borrows invisible light and changes it to red—more red light than falls on the ink actually comes from it to your eye.

**Ultraviolet bulb reveals secrets.** You can investigate all these and other tricks with borrowed light with the help of a Purple-X ultraviolet bulb, which you can buy from dealers in scientific apparatus or display lighting for about a dollar and a quarter. Basically this is just an incandescent lamp, but its extra-high-temperature filament is surrounded by a globe of red-purple glass that holds back most of the visible rays. Because most of the heat from its 250-watt filament is absorbed by the glass, the bulb should always be used in a metal reflector and should not be used more than 5 or 10 minutes at a time. When so used, its burning life should be about fifty hours.

**The "sunshine" ingredients** now used in some soaps and almost all household detergents are really colorless fluorescent dyes which attach themselves to cloth, causing the cloth to give off a blue glow when exposed to the ultraviolet rays that are always present in daylight. This glow counteracts the natural yellowness of the cloth, making it brighter than ordinary white.

**You can show how these optical bleaches work** by washing a strip of cloth in water containing a fluorescent soap (soap that glows under your ultraviolet light) and a similar strip in water containing plain soap. Even

Detergents that make cloth brighter often contain a fluorescent dye. Left strip shows effect in daylight and under ultraviolet.

under ordinary incandescent light, the strip washed with the fluorescent soap may look slightly brighter than the other (above, left). With room lights out and ultraviolet on, the difference is startling. The strip washed with the fluorescent soap shines like a brilliant bluish ghost, while the other is scarcely visible (above, right).

**Butter and colored margarine** may look alike under ordinary light, but they are usually amazingly different under ultraviolet. To find out, dissolve a little butter in several milliliters of cigarette-lighter fuel in one test tube and a little margarine in a similar amount of lighter fuel in another. When the room lights are on, both solutions appear identical. Under ultraviolet, however, the margarine fluoresces a bright blue, while the butter fluoresces a weaker yellow.

**Inks that glow under ultraviolet** but are otherwise invisible have long been tricks of spies and criminals. Such inks may be readily improvised. Prisoners have used medicines and even perspiration. A good ink of this type is a saturated solution of anthracene in benzene. Write or draw

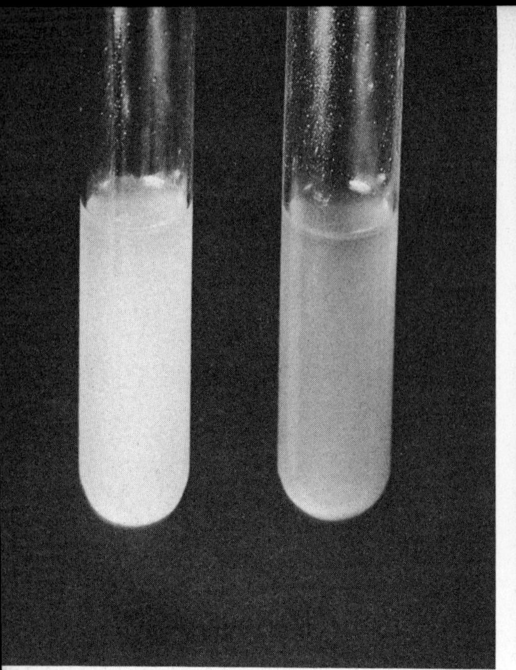

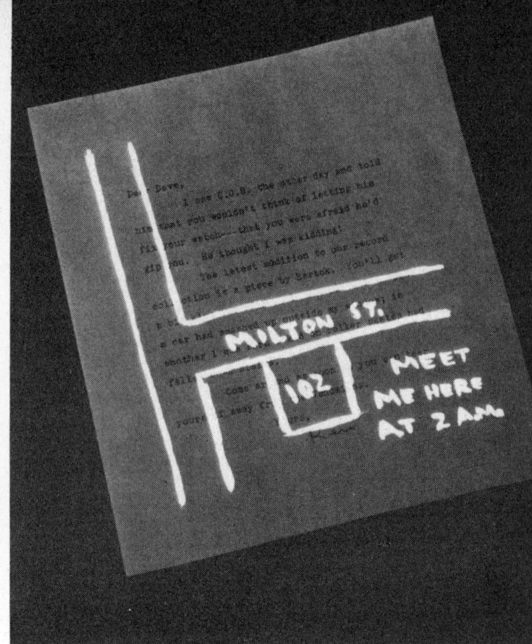


Butter or oleo? Dissolved in lighter fuel, oleo glows a bright blue, butter a weak yellow. Invisible ink shows up under UV.

with this ink on white paper and, when the writing is dry, write a message over it with ordinary ink or a typewriter. Viewed under ordinary light, only the latter message is visible, but under your "black light" the first inscription will shine brilliantly.

**Fingerprints** can often be made visible by a fluorescent method. With some powdered anthracene, you can easily show how this is done. First make a print of one of your own fingers on a piece of paper. Then sprinkle a little anthracene over the print, shake the paper to distribute the powder, and dump off the excess. Finally examine the paper under ultraviolet. Anthracene that has clung to the oil deposited from your finger will glow with a weird light.

**A test for clean dishes.** Are those restaurant and hotel dishes really clean? A simple fluorescent test devised by a chemist of the New York City Department of Health can tell in a jiffy. Suspected dishes are dipped in a solution of a fluorescent dye, rinsed, and then examined under an ultraviolet light. The dye clings to any remaining food particles. You can show how this works by dipping a dish into a cold solution of a fluorescent detergent. Let the dish drain. If there is food on it, it will shine under your Purple-X bulb.

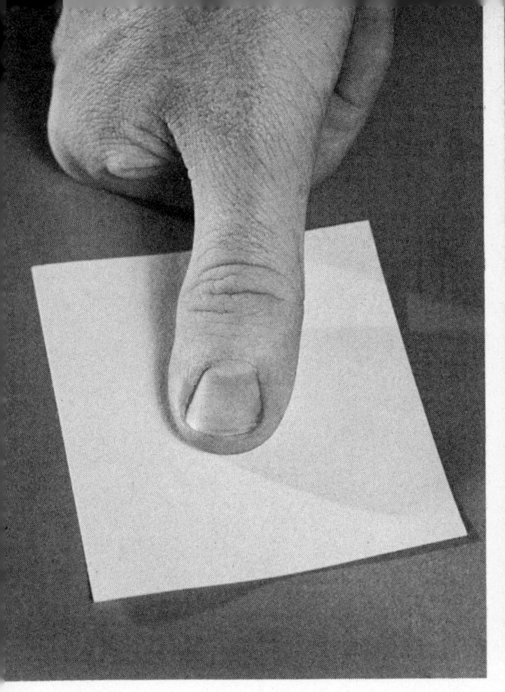

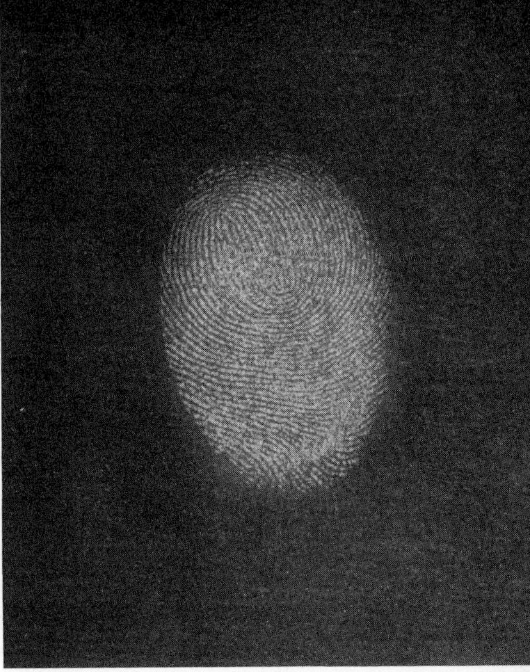

Fingerprints can be revealed by a fluorescent method. Make one and dust it with anthracene. It glows under ultraviolet light.

**Natural teeth fluoresce;** most artificial teeth and fillings do not. So you can usually find out who's got what by observing a broad smile under your ultraviolet light. If parts of some teeth shine brighter than average, you can suspect fillings of one of the new plastics.

**Fluorescent lamps** give more light per watt than incandescent bulbs, through a trick of borrowed light. The current which feeds these lamps is first converted almost completely into a narrow band of cool, ultraviolet light. A coating of fluorescent chemicals on the inside of the tube then efficiently transforms this invisible light into light you can see. The color of the light depends upon the particular chemicals. Calcium tungstate, for example, fluoresces blue; magnesium tungstate, blue-white; zinc silicate, green; zinc beryllium silicate, yellow-white.

**How these lamps change invisible to visible light** can be shown by coating half a quart jar with fluorescent paint and inverting the jar over your ultraviolet bulb. (Raise the jar on blocks to let out some of the heat, and do not operate the bulb for more than several minutes at a time.) Then turn out the room lights and switch on the ultraviolet bulb. The painted side of the jar will glow brightly, while the uncoated side will give off almost no light.

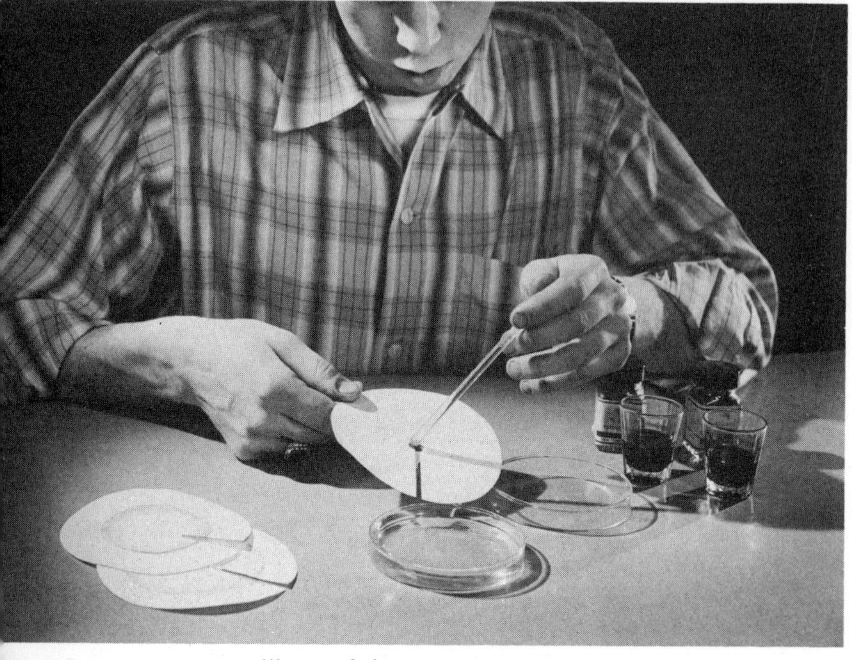

Mixtures of chemicals can often be separated in minutes by means of filter-paper disks and other simple apparatus shown here.

# Chemical Analysis by Color

BY means of bands of color, adsorbed on columns of alumina, magnesia, precipitated chalk, or even on strips of filter paper, chemists can now readily separate the constituents of complex chemical mixtures that not long ago defied division. By the same method they can also purify chemicals; concentrate vitamins, hormones, and pigments from extremely dilute solutions; identify and compare drugs, dyes, and food products almost instantly.

This "new" technique, called "adsorption chromatography" (from *chroma* meaning "color" and *graph* meaning "a writing"), was first developed in 1906 by the Russian botanist Michael Tswett, who was investigating the pigments in plant leaves. Tswett tamped precipitated chalk into a vertical glass tube, plugged at the bottom with cotton.

Then he attached this "adsorption column" to a suction flask and drew through the column an extract of dried leaf material.

**Bands of color mark different chemicals.** To Tswett's gratification the apparently uniform green pigment in the solution separated into distinct bands of different colors as it passed down the column. Tswett then forced the moist column out of the tube, cut it into sections as marked by the color bands, dissolved the pigments separately, and removed the adsorbent by filtration. What remained were solutions of pure pigments, ready for spectroscopic and chemical analysis. This pioneer experiment was the realization of a chemist's dream: the ingredients of a complex mixture were spread out for investigation like the colors of light in a spectrum. What's more, they could be cut apart with a knife!

**Substances are adsorbed at different rates.** The reason that mixtures separate on an adsorption column is essentially this: molecules of different substances travel down the column at different rates, depending upon their individual affinity to the adsorbent. Substances that have a strong affinity travel down slowly. Substances that have less affinity travel down faster. The relative affinity of different substances to the solvent also plays a part.

Tswett's simple but revolutionary method of chemical separation was little noticed for twenty-five years. In 1931 it came suddenly into prominence when the German chemists R. Kuhn and E. Lederer used it to test a solution of carotene, the yellow pigment of carrots. To their surprise, this substance, which for a hundred years had been thought to be pure, turned out to be a mixture of several compounds.

**Color analysis with paper.** From the separation of plant pigments, it was only a step to the isolation and purification of vitamins. Soon the method spread to every phase of organic and inorganic chemistry, physiology, and biology. It got a new boost when the British biochemist Archer Martin discovered that analyses could often be carried out on strips of filter paper even more easily than in an adsorption column. Today, the large-scale separation of chemicals is performed in adsorption columns, while analysis of small quantities is generally done on paper.

Although ingenuity is often required in selecting the best solvent for a particular mixture and in making colorless substances visible, the basic technique of paper chromatography is simple. In one method, a line is drawn with a solution of the substance to be tested near one end of a

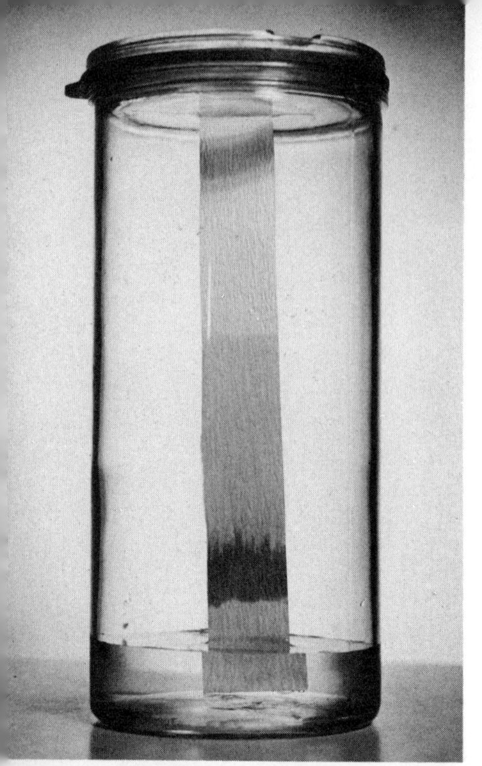

In strip analysis, a long strip of filter paper is supported in a jar, its lower end dipping in a solvent. A stripe of chemical to be analyzed is made just above the solvent. As the solvent rises, it separates the mixture.

long strip of filter paper. The strip is then suspended in a closed transparent container so that this end dips, not quite to the line, in a suitable solvent. By capillary action, the solvent creeps up through the questioned substance, dissolves it, and redeposits it along the strip. By confining the solvent vapor, the container prevents the solvent from evaporating from the strip.

Some substances will separate out completely on a short strip of paper; others require strips up to 400 mm long.

**You don't need elaborate equipment** to demonstrate how this amazing method works. A quart glass refrigerator jar and a few strips of filter paper, about 25 mm wide and 200 mm long, will do for apparatus. Household inks and dyes will serve as substances to be tested, and water can be the first solvent.

Bend at right angles about 20 mm of one end of a filter paper strip and attach this "leg" to the under side of the jar cover with cellulose tape. Next paint a narrow stripe across the strip with washable ink, a food color, or a water-soluble dye, about 30 mm from the other end. Put enough water in the jar to reach up about 15 mm on the strip when the cover is put on with the strip suspended from it. As soon as the color dries, put on the cover.

**Purple separates into red and blue.** The water slowly climbs the paper. When it reaches the colored stripe, it dissolves the color and carries it along. If the original ink or dye is a pure chemical substance, its color will not change. If it is a mixture, however, the different components will gradu-

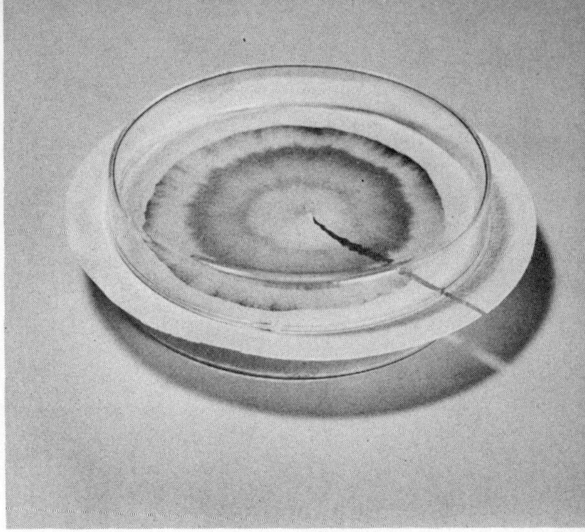

Make the "wick" for disk analysis as shown left. This dips in solvent in the lower petri dish. A second dish makes a cover.

ally separate. Purple dye, for example, may amazingly separate into red and blue; green may divide into blue and yellow. The farther the solvent continues up the strip, the greater will be the separation.

**Chromatography with filter-paper disks.** In another simple technique, a disk of filter paper, 11 or 12½ cm in diameter, is used in place of a strip. A drop of a solution of the substance to be tested is placed on the center of the disk, the disk is sandwiched between the bottom and cover of a 100-mm petri dish (or better, between two bottoms or two covers), and the solvent is fed from the bottom dish to the center of the disk through a wick. As the solvent advances, the substance spreads across the paper in ever-widening circles.

This method is convenient and rapid. To try it, first prepare a disk by making two parallel scissor cuts, about 3 mm apart, from the edge of the disk to the center. Then bend the narrow strip thus formed at right angles to the disk and cut it so the remaining stub is about 15 mm long. This stub is the wick.

Place a dot of the substance to be tested on the juncture of the disk and the wick. As soon as the dot has dried, lay the disk on the lower part of a petri dish containing several milliliters of solvent. After making sure the wick dips in the solvent, set the upper part of the dish on the disk. Substances are sometimes clearly separated in a few minutes. Separation becomes greater as the bands of color move toward the edge.

**Solvent determines spread of colors.** With a given paper and solvent, the arrangement of the bands in the chromatogram of a particular sub-

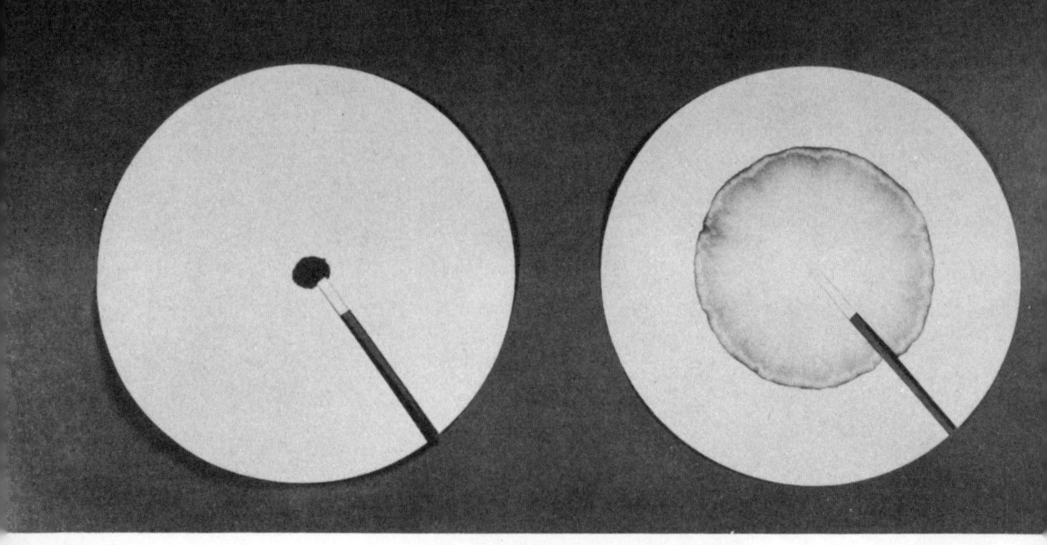

Spot of purple dye in center of disk, at left, is spread out into separate rings of red and blue as it is carried along by solvent.

stance is as constant as the fingerprint of an individual. Often, however, the use of a different solvent will change the arrangement completely. The expert analyst makes use of this difference to spread out substances that ordinarily huddle together, or even to reverse the order of the bands.

You can demonstrate such an "about-face" by making two chromatograms of a solution containing a mixture of Victoria blue and methylene blue dyes—one using a solvent of plain water and the other a solvent consisting of a mixture of acetone (*flammable!*) with 5 parts water and 8 parts hydrochloric acid. In the first chromatogram, methylene blue leads with Victoria close behind; in the second, Victoria blue comes out far ahead.

**Colorless substances made visible.** Many ordinarily colorless substances can be separated by paper chromatography and then made visible by streaking the chromatogram with special reagents, applied with a small brush. For example, make a chromatogram of a dilute solution of cobalt chloride and ferrous sulfate in water, using the acetone, water, and hydrochloric acid mixture mentioned above as the spreading solvent. The bands produced by the two chemicals are almost invisible. When dry, paint a streak across them with a reagent consisting of equal parts of a saturated solution of potassium thiocyanate and acetone. The ferric band will turn blood-red, the cobalt band bright blue.

**Black light identifies bands.** Other colorless chemicals can often be revealed and identified by viewing a chromatogram under ultraviolet

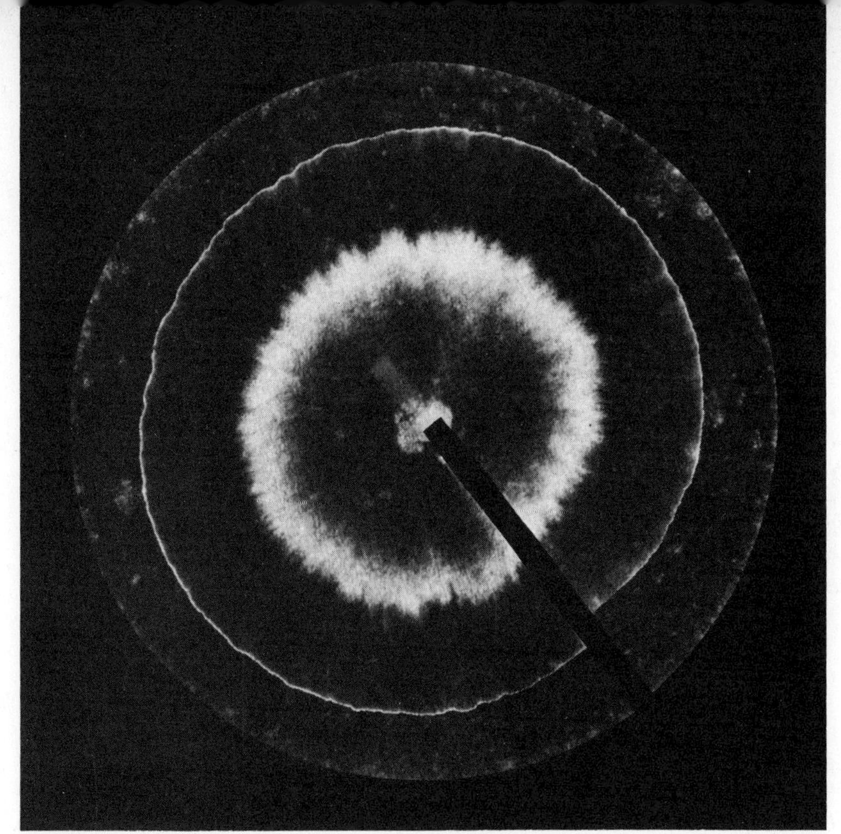

**Bands of ordinarily colorless lemon extract and riboflavin shine out as blue and brilliant yellow-green under ultraviolet light.**

light. Lemon extract, caffeine in coffee, the colorless dye in some synthetic detergents, certain vitamins, and many other common substances will give a tell-tale glow under these invisible rays.

For a vivid demonstration, make a chromatogram of a mixture of lemon extract with an equal amount of a solution of riboflavin (vitamin B-2), using water as the spreading solvent. (To make the riboflavin solution, first wash off the outer coating of a 5-mg tablet, then dissolve the inside in about 50 ml of water.) Under ordinary light this chromatogram looks practically blank. But turn out the room lights and turn on the "black light" from a Purple-X bulb (see page 162). The band of lemon extract shines out with a pale-blue light, that of the riboflavin with a brilliant yellow-green!

# USEFUL FACTS AND FORMULAS

## IMPORTANT COMMON CHEMICALS

| Common name | Chemical name | Formula |
|---|---|---|
| Alcohol | Ethyl alcohol or ethanol | $C_2H_5OH$ |
| Alum | Aluminum potassium sulfate | $AlK(SO_4)_2 \cdot 12H_2O$ |
| Alumina | Aluminum oxide | $Al_2O_3$ |
| Aqua ammonia | Ammonium hydroxide solution | $NH_4OH + H_2O$ |
| Aqua fortis | Nitric acid | $HNO_3$ |
| Aqua regia | Nitric and hydrochloric acids | $HNO_3 + HCl$ |
| Aspirin | Acetylsalicylic acid | $C_2H_3O_2C_6H_4CO_2H$ |
| Baking soda | Sodium bicarbonate | $NaHCO_3$ |
| Banana oil | Amyl acetate | $CH_3CO_2C_5H_{11}$ |
| Baryta | Barium oxide | $BaO$ |
| Benzol | Benzene | $C_6H_6$ |
| Bichloride of mercury | Mercuric chloride | $HgCl_2$ |
| Black lead | Graphite (mineral carbon) | $C$ |
| Black oxide of copper | Cupric oxide | $CuO$ |
| Black oxide of mercury | Mercurous oxide | $Hg_2O$ |
| Bleaching powder. | Calcium hypochlorite | $CaOCl_2$ |
| Blue vitriol | Copper sulfate | $CuSO_4 \cdot 5H_2O$ |
| Boracic acid | Boric acid | $H_3BO_3$ |
| Borax | Sodium borate | $Na_2B_4O_7 \cdot 10H_2O$ |
| Brimstone | Sulfur | $S$ |
| "Butter of" | Chloride or trichloride of | |
| Calomel | Mercurous chloride | $Hg_2Cl_2$ |
| Carbolic acid | Phenol | $C_6H_5OH$ |
| Carbonic acid | Carbon dioxide | $CO_2$ |
| Caustic potash | Potassium hydroxide | $KOH$ |
| Caustic soda | Sodium hydroxide | $NaOH$ |
| Chalk | Calcium carbonate | $CaCO_3$ |
| Chile saltpeter | Sodium nitrate | $NaNO_3$ |
| Chloride of lime | Calcium hypochlorite | $CaOCl_2$ |
| Chloroform | Trichloromethane | $CHCl_3$ |
| Chrome alum | Chromium potassium sulfate | $CrK(SO_4)_2 \cdot 12H_2O$ |
| Copperas | Ferrous sulfate | $FeSO_4 \cdot 7H_2O$ |
| Corrosive sublimate | Mercuric chloride | $HgCl_2$ |
| Cream of Tartar | Potassium bitartrate | $KHC_4H_4O_6$ |
| Crocus powder | Ferric oxide | $Fe_2O_3$ |
| Dry ice | Solid carbon dioxide | $CO_2$ |
| Dutch liquid | Ethylene dichloride | $(CH_2Cl)_2$ |
| Emery powder | Impure aluminum oxide | $Al_2O_3$ |
| Epsom salts | Magnesium sulfate | $MgSO_4 \cdot 7H_2O$ |
| Ether | Ethyl ether | $(C_2H_5)_2O$ |
| Fluorspar | Natural calcium fluoride | $CaF_2$ |

## IMPORTANT COMMON CHEMICALS (*Continued*)

| Common name | Chemical name | Formula |
|---|---|---|
| French chalk | Natural magnesium silicate | $H_2Mg_3(SiO_3)_4$ |
| Galena | Natural lead silfide | $PbS$ |
| Glauber's salt | Sodium sulfate | $Na_2SO_4 \cdot 10H_2O$ |
| Grain alcohol | Ethyl alcohol or ethanol | $C_2H_5OH$ |
| Green vitriol | Ferrous sulfate | $FeSO_4 \cdot 7H_2O$ |
| Gypsum | Natural calcium sulfate | $CaSO_4 \cdot 2H_2O$ |
| Hypo | Sodium thiosulfate | $Na_2S_2O_3 \cdot 5H_2O$ |
| Javelle water | Sodium hypochlorite solution | $NaOCl + H_2O$ |
| Laughing gas | Nitrous oxide | $N_2O$ |
| Lime | Calcium oxide | $CaO$ |
| Limewater | Calcium hydroxide solution | $Ca(OH)_2+H_2O$ |
| Litharge | Lead oxide | $PbO$ |
| Lunar caustic | Silver nitrate | $AgNO_3$ |
| Lye | Sodium hydroxide | $NaOH$ |
| Magnesia | Magnesium oxide | $MgO$ |
| Marble | Calcium carbonate | $CaCO_3$ |
| Methylated spirits | Methyl alcohol or methanol | $CH_2OH$ |
| Milk of magnesia | Magnesium hydroxide in water | $Mg(OH)_2$ |
| Minium | Lead tetroxide | $Pb_3O_4$ |
| "Muriate of" | Chloride of | |
| Muriatic acid | Hydrochloric acid | $HCl$ |
| Niter | Potassium nitrate | $KNO_3$ |
| Oil of bitter almonds (artificial) | Benzaldehyde | $C_6H_5CHO$ |
| Oil of mirbane | Nitrobenzene | $C_6H_5NO_2$ |
| Oil of vitriol | Sulfuric acid | $H_2SO_4$ |
| Oil of wintergreen (artificial) | Methyl salicylate | $C_6H_4OHCOOCH_3$ |
| Oleum | Fuming sulfuric acid | $H_2SO_4SO_3$ |
| Pearl ash | Potassium carbonate | $K_2CO_3$ |
| Peroxide | Peroxide of hydrogen solution | $H_2O_2 + H_2O$ |
| Plaster of paris | Calcium sulfate | $(CaSO_4)_2 \cdot H_2O$ |
| Plumbago | Graphite (mineral carbon) | $C$ |
| Potash | Potassium carbonate | $K_2CO_3$ |
| Prussic acid | Hydrocyanic acid | $HCN$ |
| Pyro | Pyrogallic acid | $C_6H_3(OH)_3$ |
| Quicklime | Calcium oxide | $CaO$ |
| Quicksilver | Mercury | $Hg$ |
| Red lead | Lead tetroxide | $Pb_3O_4$ |
| Red oxide of copper | Cuprous oxide | $Cu_2O$ |
| Red oxide of mercury | Mercuric oxide | $HgO$ |
| Rochelle salt | Potassium sodium tartrate | $KNaC_4H_4O_6 \cdot 4H_2O$ |
| Rouge | Ferric oxide | $Fe_2O_3$ |

## IMPORTANT COMMON CHEMICALS (*Continued*)

| Common name | Chemical name | Formula |
|---|---|---|
| Sal ammoniac | Ammonium chloride | $NH_4Cl$ |
| Saleratus | Sodium bicarbonate | $NaHCO_3$ |
| Sal soda | Crystalline sodium carbonate | $Na_2CO_3 \cdot 10H_2O$ |
| Salt | Sodium chloride | $NaCl$ |
| Saltpeter | Potassium nitrate | $KNO_3$ |
| Salts of lemon | Potassium binoxalate | $KHC_2O_4 \cdot H_2O$ |
| Silica | Silicon dioxide | $SiO_2$ |
| Slaked lime | Calcium hydroxide | $Ca(OH)_2$ |
| Soda ash | Dry sodium carbonate | $Na_2CO_3$ |
| Spirits of salts | Hydrochloric acid | $HCl$ |
| Spirits of wine | Ethyl alcohol or ethanol | $C_2H_5OH$ |
| Sugar of lead | Lead acetate | $Pb(C_2H_3O_2)_2 \cdot 3H_2O$ |
| Sulfuric ether | Ethyl ether | $(C_2H_5)_2O$ |
| Talc | Magnesium silicate | $H_2Mg_3(SiO_3)_4$ |
| Vinegar | Dilute and impure acetic acid | $CH_3COOH$ |
| Washing soda | Crystalline sodium carbonate | $Na_2CO_3 \cdot 10H_2O$ |
| Water glass | Sodium silicate | $Na_2SiO_3$ |
| White arsenic | Arsenic trioxide | $As_2O_3$ |
| White lead | Basic lead carbonate | $(PbCO_3)_2 \cdot Pb(OH)_2$ |
| White vitriol | Zinc sulfate | $ZnSO_4 \cdot 7H_2O$ |
| Whiting | Powdered calcium carbonate | $CaCO_3$ |
| Wood alcohol | Methyl alcohol or methanol | $CH_3OH$ |
| Zinc white | Zinc oxide | $ZnO$ |

# WEIGHTS AND MEASURES

## RELATIONSHIP BETWEEN ENGLISH AND METRIC UNITS

1 meter (m) = 39.37 inches
1 inch = 2.54 centimeters (cm)
1 liter (l) = 1.06 quarts
1 kilogram (kg) = 2.2 pounds

1 ounce (avoirdupois) = 28.3 grams (g)
1 pound = 453.6 grams (g)
1 gallon (U.S.) = 3.8 liters (l)
1 ounce (fluid, U.S.) = 29.57 milliliters (ml)

## MISCELLANEOUS EQUIVALENTS

1 cubic foot of water = 62.4 pounds
1 gallon of water = 8.35 pounds
1 drop of water = 1/20 milliliter (ml)

## EMERGENCY WEIGHING WITH COINS

A dime weighs approximately 2.5 grams
A cent weighs 3 grams
A nickel weighs 5 grams
A quarter weighs 6.25 grams

## SPECIFIC GRAVITY OF COMMON SUBSTANCES

| Substance | Specific gravity | Substance | Specific gravity |
|---|---|---|---|
| Cork | 0.24 | Chloroform | 1.489 |
| Gasoline | 0.66–0.69 | Magnesium | 1.74 |
| Ether | 0.763 | Aluminum | 2.7 |
| Alcohol | 0.791 | Iron | 7.9 |
| Water | 1.00 | Lead | 11.3 |

## TEMPERATURES USEFUL TO KNOW

| Centigrade | Fahrenheit | |
|---|---|---|
| —273 | —459.4 | Absolute zero |
| —130 | —202 | Alcohol freezes |
| —78.5 | —109.3 | Dry ice sublimes |
| —38.9 | —38 | Mercury freezes |
| 0 | 32 | Ice melts |
| 34.5 | 94.1 | Ether boils |
| 37 | 98.4 | Temperature of human body |
| 60 | 140 | Wood's metal melts |
| 78.5 | 173.3 | Alcohol boils |
| 100 | 212 | Water boils |
| 160 | 320 | Sugar melts |
| 232 | 450 | Tin melts |
| 327 | 621 | Lead melts |
| 658 | 1,216 | Aluminum melts |
| 700 | 1,292 | Dull red heat |
| 800 | 1,472 | Pyrex glass begins to soften |
| 1,000 | 1,832 | Bright red heat |
| 1,083 | 1,980 | Copper melts |
| 1,400 | 2,552 | White heat |
| 1,500 | 2,732 | Temperature of bunsen flame |
| 1,530 | 2,786 | Iron melts |
| 1,773 | 3,223 | Platinum melts |
| 4,000 | 7,232 | Temperature of electric furnace |
| 6,000 | 10,800 | Temperature of sun's surface |

### TEMPERATURE CONVERSION

Centigrade to Fahrenheit: Multiply the centigrade reading by 9/5 and add 32.

Fahrenheit to centigrade: Subtract 32 from the Fahrenheit reading and multiply by 5/9.

# Index